Yearbook of Astronomy 2022

Front Cover: This iconic image of the Southern Region of the Moon was captured on 15 September 1919 by astronomer Francis Gladheim Pease through the 100-inch Hooker telescope at Mount Wilson Observatory, California. This was the image used on the cover of the *Yearbook of Astronomy 1962*. For further details see the article 'The Yearbook of Astronomy Cover Image: 1962 and 2022' by Steve Brown on page 216 of this edition. (Image courtesy Carnegie Institution for Science)

YEARBOOK OF
ASTRONOMY
2022

EDITED BY

Brian Jones

WHITE OWL
AN IMPRINT OF PEN & SWORD BOOKS LTD.
YORKSHIRE · PHILADELPHIA

First published in Great Britain in 2021 by
WHITE OWL
An imprint of
Pen & Sword Books Ltd
Yorkshire – Philadelphia

ISBN 978 1 52679 005 7

Typeset in Dante By Mac Style

Printed and Bound in India by Replika Press Pvt. Ltd.

Pen & Sword Books Ltd incorporates the Imprints of Pen & Sword Books Archaeology, Atlas, Aviation, Battleground, Discovery, Family History, History, Maritime, Military, Naval, Politics, Railways, Select, Transport, True Crime, Fiction, Frontline Books, Leo Cooper, Praetorian Press, Seaforth Publishing, Wharncliffe and White Owl.

For a complete list of Pen & Sword titles please contact

PEN & SWORD BOOKS LIMITED
47 Church Street, Barnsley, South Yorkshire, S70 2AS, England
E-mail: enquiries@pen-and-sword.co.uk
Website: www.pen-and-sword.co.uk

or

PEN AND SWORD BOOKS
1950 Lawrence Rd, Havertown, PA 19083, USA
E-mail: Uspen-and-sword@casematepublishers.com

Contents

Editor's Foreword	8
Preface	14
About Time	15
Using the Yearbook of Astronomy as an Observing Guide	17

The Monthly Star Charts
Northern Hemisphere Star Charts *David Harper*	27
Southern Hemisphere Star Charts *David Harper*	53

The Planets in 2022 *Lynne Marie Stockman*	78
Some Events in 2022	81
Phases of the Moon in 2022	84
Eclipses in 2022	85

Monthly Sky Notes and Articles 2022
Monthly Sky Notes January	88
Morning Apparition of Venus – January 2022 to October 2022	88
Mars – January 2022 to June 2022	90
Joseph Ward: New Zealand Polymath *John McCue*	92

Monthly Sky Notes February	97
Jacobus Cornelius Kapteyn *David M. Harland*	99

Monthly Sky Notes March	103
A True Pioneer of Planetary Exploration *Neil Haggath*	105

Monthly Sky Notes April	108
Biela's Comet: A Tale of Two Parts *Neil Norman*	111

Monthly Sky Notes May	115
Maximilian Hell: A Legacy in Transit *Richard Hill*	117

Monthly Sky Notes June   122
Margherita Hack *Mary McIntyre*   124

Monthly Sky Notes July   127
Gravity Assists: Something for Nothing? *Peter Rea*   129

Monthly Sky Notes August   134
Saturn – January 2022 to December 2022   135
U.S. Mints Celestial Themes *Carolyn Kennett*   137

Monthly Sky Notes September   140
Jupiter – January 2022 to December 2022   141
Neptune – January 2022 to December 2022   142
Early Astronomy Stamps from Brazil *Katrin Raynor-Evans*   144

Monthly Sky Notes October   147
Evening Apparition of Venus – October 2022 to August 2023   148
Mills Observatory *Katrin Raynor-Evans*   150

Monthly Sky Notes November   153
Uranus in 2022   154
The Hoba West Meteorite *Susan Stubbs*   156

Monthly Sky Notes December   159
Mars – July 2022 to January 2023   160
The Next 'Small Step' *David M. Harland*   162

Comets in 2022 *Neil Norman*   166
Minor Planets in 2022 *Neil Norman*   173
Meteor Showers in 2022 *Neil Norman*   181

Article Section
Astronomy in 2021 *Rod Hine*   191
Solar System Exploration in 2021 *Peter Rea*   199
Anniversaries in 2022 *Neil Haggath*   208
The Yearbook of Astronomy Cover Image: 1962 and 2022 *Steve Brown*   216
A History of the Amateur Astronomical Society: 1962 to 2022
   Allan Chapman   218

Expanding Cosmic Horizons *Martin Rees*   227

The Astronomers' Stars: A Study in Scarlet *Lynne Marie Stockman*   238

Frank Drake and His Equation *David M. Harland*   249

Remote Observing and Imaging *Damian Peach*   261

Skies Over Ancient America *P. Clay Sherrod*   270

Tycho Brahe and the Parallax of Mars *David Harper*   280

A Tale of Two Henrys and the Search for Their Great Telescopes *Gary Yule*   286

Mission to Mars: Countdown to Building a Brave New World: Laying the
    Foundations *Martin Braddock*   293

Ad Astra: A Personal Journey *David H. Levy*   300

Internet Satellites: Less Welcome Constellations? *Bob Mizon*   308

Miscellaneous

Some Interesting Variable Stars *Tracie Heywood*   317

Some Interesting Double Stars *Brian Jones*   330

Some Interesting Nebulae, Clusters and Galaxies *Brian Jones*   333

Astronomical Organizations   335

Our Contributors   340

Sky & Telescope (Advertisement)   347

Yearbook of Astronomy 2017 (Advertisement)   349

Editor's Foreword

The *Yearbook of Astronomy 2022* is the Diamond Jubilee edition of this iconic publication, the annual appearance of which has been eagerly anticipated by astronomers, both amateur and professional, ever since this invaluable book first appeared in 1962. As the preface to the *Yearbook of Astronomy 1962* informed its readers, the post-war years had seen a tremendous growth of interest in astronomy and space research. The launch of Sputnik 1 in October 1957 heralded the dawn of the Space Age and a significant change in the course of history. This, coupled with the subsequent flights of Soviet cosmonaut Yuri Gagarin (April 1961) and American astronaut Alan Shepard (May 1961), served to engender a public interest in astronomy and space that has continued to grow and expand to this day.

Prior to the *Yearbook of Astronomy*, there was no publication available to fulfil the ever-growing popular demand for up-to-date information on astronomy and space exploration. The aim and intent of the *Yearbook* was to meet this need, and it has successfully continued to inform and educate its readers ever since those far-flung days of 1962.

Aimed at both the armchair astronomer and the active backyard observer, the *Yearbook of Astronomy 2022* continues the trend of previous editions. Within its pages you will find a rich blend of information, star charts and guides to the night sky coupled with an interesting mixture of articles which collectively embrace a wide range of topics, ranging from the history of astronomy to the latest results of astronomical research; space exploration to observational astronomy; and our own celestial neighbourhood out to the farthest reaches of space.

The *Monthly Star Charts* have been compiled by David Harper and show the night sky as seen throughout the year. Two sets of twelve charts have been provided, one set for observers in the Northern Hemisphere and one for those in the Southern Hemisphere. Between them, each pair of charts depicts the entire sky as two semi-circular half-sky views, one looking north and the other looking south.

Lists of *Phases of the Moon in 2022* and *Eclipses in 2022* are also provided, together with general summaries of the observing conditions for each of the planets in *The Planets in 2022*, and a calendar of significant Solar System events occurring throughout the year in *Some Events in 2022*.

The ongoing process of improving and updating what the *Yearbook of Astronomy* offers to its readers is continued in the 2022 edition with apparition charts, compiled by David Harper, for all the major planetary members of our Solar System. Further details of the planetary apparition charts are given in the article *Using the Yearbook of Astronomy as an Observing Guide*.

As with *The Planets in 2022* and *Some Events in 2022*, the *Monthly Sky Notes* have been compiled by Lynne Marie Stockman and give details of the positions and visibility of the planets for each month throughout 2022. Each section of the *Monthly Sky Notes* is accompanied by a short article, the range of which includes items on a variety of astronomy-related topics such as *Joseph Ward: New Zealand Polymath* by John McCue; *Maximilian Hell: A Legacy in Transit* by Richard Hill; and *Gravity Assists: Something for Nothing?* by Peter Rea.

The Monthly Sky Notes and Articles section of the book concludes with a trio of articles penned by Neil Norman, these being *Comets in 2022*, *Minor Planets in 2022* and *Meteor Showers in 2022*, all three titles being fairly self-explanatory describing as they do the occurrence and visibility of examples of these three classes of object during and throughout the year.

In his article *Astronomy in 2021* Rod Hine discovers that the origin of the heavier metals, especially gold, may be more complicated than the simplistic explanation that "elements heavier than iron were forged in supernova". Meticulous research and modelling suggests that there are other mechanisms, some perhaps still undiscovered. Other topics covered include two new observatories, one about to come on stream and another just flexing its muscles, both of which will provide fast and comprehensive data for future researchers. There is also a short tribute to the Arecibo Radio Telescope whose collapse in late-2020 brought to a close an amazing five decade long chapter of scientific success.

This is followed by *Solar System Exploration in 2021* in which Peter Rea updates us on the progress of a number of planetary missions, including HOPE (the United Arab Emirates first interplanetary mission to Mars) and China's First Mars Mission Tianwen-1. Also covered are the two asteroid sample collection and return flights of the Japanese Hayabusa2 (162173 Ryugu) and the American OSIRIS-Rex (101955 Bennu) spacecraft. Hayabusa2 returned samples in December 2020 with OSIRIS-Rex due to return to Earth with its precious cargo in September 2023.

In 2022 we celebrate the 450th anniversary of the appearance of a supernova in the constellation Cassiopeia, one of only eight supernovae visible to the naked eye in recorded history. Reaching its peak brightness on 16 November 1572, this object became known as Tycho's Star, the most accurate observations of the phenomenon having being made by the great Danish astronomer Tycho Brahe. Neil Haggath discusses this event in his article *Anniversaries in 2022* as well as drawing our attention to the bicentenary of the death of William Herschel (1738–1822), widely regarded as the greatest visual observational astronomer who ever lived, and the 50th anniversary of the death of Harlow Shapley (1885–1972), one of the pioneers of modern astrophysics. We are also reminded that 50 years ago, in December 1972, what was arguably humanity's greatest adventure came to an end when Apollo 17 mission Commander Gene Cernan became the last person, so far, to walk on the surface of the Moon.

The short article *The Yearbook of Astronomy Cover Image: 1962 and 2022*, penned by regular contributor Steve Brown, draws comparisons between the cover designs from the 1962 and 2022 editions of the Yearbook, as well as describing the subject matter of the photograph and briefly outlining the original source of what is held to be an iconic image by many readers of the *Yearbook of Astronomy*. The article includes an illustration depicting the cover design of the 1962 edition. Every effort was made to contact the copyright holders of the early editions, unfortunately with little success.

The amateur astronomical society was a Victorian creation, with membership typically drawn from the ranks of urban middle-class professionals with scarcely a working man or woman from the industrial districts. However, as we learn in the fascinating article *A History of the Amateur Astronomical Society: 1962 to 2022* by Allan Chapman, this situation had changed dramatically by 1962 with the dawn of the Space Age (and the first appearance of the *Yearbook of Astronomy*). All this occurred at a time when the growing accessibility of radio and television had brought a whole realm of knowledge straight into everyone's living room, transforming the leisure time enjoyed by working people, and inspiring them to start meeting informally together to discuss the new astronomical fascination. A new era had begun for astronomical societies, and in this article we see how their role has changed and evolved over the last six decades, and are reassured to learn that astronomical societies are just as strong in 2022 as they were back in 1962.

When the history of recent science is eventually written, one of the most important chapters will describe how astronomers have discovered a great deal about cosmic history. Over a period of 13.8 billion years, our universe has evolved from a hot dense beginning into the panorama of galaxies, stars and planets that we see around us today … and within which life has emerged. In his article *Expanding Cosmic Horizons*, Martin Rees describes some key steps in our understanding – mainly due to improved instruments on the ground and in space – and ends by raising some of the more speculative questions that challenge us today.

The Astronomers' Stars: A Study in Scarlet by Lynne Marie Stockman is the first in a series of articles exploring remarkable and unusual stars named after the equally remarkable astronomers who discovered, investigated or defined them. Red stars have fascinated astronomers since the eighteenth century. Observers noticed that many of the variable stars they watched were tinged with red and some astronomers thought that red stars formed a distinct class of stellar object, worthy of serious attention. The article reviews William Herschel's Garnet Star, his son John Herschel's Ruby Star and Hind's Crimson Star, discussing what makes these men and their stars special.

In the article *Frank Drake and His Equation* regular contributor David M. Harland looks back at the work of young radio astronomer Frank Drake of the National Radio Astronomy Observatory in America who, in 1962, coordinated the first systematic attempt to calculate the probability of there being alien civilisations in the galaxy. The subsequent discovery of planetary systems around thousands of stars has shed new light on the perceived limitations of Drake's methodology.

The light-polluted skies under which many backyard astronomers and aspiring astrophotographers observe can often restrict their efforts to capture that perfect image. However, in the article *Remote Observing and Imaging* by Damian Peach, we are introduced to the concept of remote astronomy, and the availability of remote telescopes through which we can now observe and image the night sky from a range of observatories scattered across the globe. As we discover, remote telescopes have revolutionised astrophotography, and we are now able to access and use professional telescopes at remote observatories – through our own computers at home – to capture that once-only-dreamed-of image of a celestial object that was either too faint to be seen through our own telescope, or was visually-inaccessible due to light pollution.

Following on from articles relating to Aboriginal astronomy and Maori astronomy in recent editions of the Yearbook, we have *Skies Over Ancient America* by P. Clay Sherrod, the first in a series of three articles describing the evolution of astronomical understanding across North America. Long ago – so the questions can never be fully answered – the prehistoric people of North America developed a desire, and later a need, for detailed study of the heavens above: the sun, moon and stars. Myths and legends relating to these objects still remain today, and throughout the North American Continent remain remarkable structures that signify the sky as a critical tool for survival, and even cultural advancement. We discover where it all began, and how this earliest astronomy evolved as Native American people migrated while following a changing climate.

In *Tycho Brahe and the Parallax of Mars* David Harper examines the Danish astronomer's most famous observations, which were later used by Johannes Kepler to establish his laws of planetary motion. Tycho is rightly celebrated as the greatest observational astronomer of the pre-telescopic era, but the tale of the parallax of Mars demonstrates that he also deserves recognition as a theoretical cosmologist.

A Tale of Two Henrys and the Search for Their Great Telescopes by Gary Yule tells us of his efforts to track down the current locations of two famous telescopes. The instruments featured in the search were the eleven-foot Dollond refractor owned by Henry Lawson (1774–1855), and the twelve-inch Truss Tube Newtonian Reflector which belonged to Henry Brinton (1901–1977) and which graced the front cover of the *Yearbook of Astronomy 1963*. As we discover, Gary's searches produced mixed results.

Mission to Mars: Countdown to Building a Brave New World: Laying the Foundations by Martin Braddock is the second in a series of articles scheduled to appear in the Yearbook of Astronomy throughout the 2020s and which will keep the reader fully up to date with the ongoing preparations geared towards sending a manned mission to Mars at or around the turn of the decade.

Ad Astra: A Personal Journey by David H. Levy captures the life of an accomplished astronomer, science writer and comet-hunter in just a few fascinating and highly-readable pages. The article is a review of how David first became interested in the night sky – at age 12, in 1960 – and how five years later began a search for comets that continues to this day. It is the writer's hope that readers will enjoy following the successes and failures of that life, and that in some way they may also be inspired to reach for the stars.

In his article *Internet Satellites: Less Welcome Constellations* Bob Mizon describes the threat to astronomy and especially astrophotography of the thousands of global internet satellites, launched with little consultation with appropriate agencies, but which are filling our ancient heritage of dark night skies with moving points of light. Connection for all – but at what price?

The final section of the book starts off with *Some Interesting Variable Stars* by Tracie Heywood which contains useful information on variables as well as predictions for timings of minimum brightness of the famous eclipsing binary Algol for 2022. *Some Interesting Double Stars* and *Some Interesting Nebulae, Star Clusters and Galaxies* present a selection of objects for you to seek out in the night sky. The lists included here are by no means definitive and may well omit your favourite celestial targets. If this is the case, please let us know and we will endeavour to include these in future editions of the *Yearbook*.

The book rounds off with a selection of *Astronomical Organizations*, which lists organizations and associations across the world through which you can further pursue your interest and participation in astronomy (if there are any that we have omitted please let us know) and *Our Contributors*, which contains brief background details of the numerous writers who have contributed to this edition of the

Yearbook. Finally, regular readers of the Yearbook will notice that this edition does not include a Glossary. This has now been transferred to the *Yearbook of Astronomy* website at **www.yearbookofastronomy.com** and readers of the Yearbook will be invited to submit entries for inclusion in the Glossary, further details of which will be announced in due course.

Over time new topics and themes will be introduced into the Yearbook to allow it to keep pace with the increasing range of skills, techniques and observing methods now open to amateur astronomers, this in addition to articles relating to our rapidly-expanding knowledge of the Universe in which we live. There will be an interesting mix, some articles written at a level which will appeal to the casual reader and some of what may be loosely described as at a more academic level. The intention is to fully maintain and continually increase the usefulness and relevance of the Yearbook of Astronomy to the interests of the readership who are, without doubt, the most important aspect of the Yearbook and the reason it exists in the first place. With this in mind, suggestions from readers for further improvements and additions to the Yearbook content are always welcomed. After all, the book is written for you ...

As ever, grateful thanks are extended to those individuals who have contributed a great deal of time and effort to the *Yearbook of Astronomy 2022*, including David Harper, who has provided updated versions of his excellent Monthly Star Charts. These were generated specifically for what has been described as the new generation of the *Yearbook of Astronomy*, and the charts add greatly to the overall value of the book to star gazers. Equally important are the efforts of Lynne Marie Stockman who has put together the Monthly Sky Notes. Their combined efforts have produced what can justifiably be described as the backbone of the *Yearbook of Astronomy*. My grateful thanks are also due to Harold A. McAlister, Director Emeritus, Mount Wilson Observatory and Kit Whitten, Library Assistant/ Archivist at Carnegie Science for their help with sourcing and providing the image on the front cover, and to Rod Hine for his valuable help in sourcing several images used in this edition. Also worthy of mention are Mat Blurton, who has done an excellent job typesetting the Yearbook, and Jonathan Wright, Emily Robinson, Lori Jones, Janet Brookes, Paul Wilkinson, Quinn Padgett, Charlie Simpson and Rosie Crofts of Pen & Sword Books Ltd for their efforts in producing and promoting the *Yearbook of Astronomy 2022*, the Diamond Jubilee edition of this much-loved and iconic publication.

Brian Jones – Editor
Bradford, West Riding of Yorkshire
January 2021

Preface

The information given in this edition of the Yearbook of Astronomy is in narrative form. The positions of the planets given in the Monthly Sky Notes often refer to the constellations in which they lie at the time. These can be found on the star charts which collectively show the whole sky via two charts depicting the northern and southern circumpolar stars and forty-eight charts depicting the main stars and constellations for each month of the year. The northern and southern circumpolar charts show the stars that are within 45° of the two celestial poles, while the monthly charts depict the stars and constellations that are visible throughout the year from Europe and North America or from Australia and New Zealand. The monthly charts overlap the circumpolar charts. Wherever you are on the Earth, you will be able to locate and identify the stars depicted on the appropriate areas of the chart(s).

There are numerous star atlases available that offer more detailed information, such as *Sky & Telescope's POCKET SKY ATLAS* and *Norton's STAR ATLAS and Reference Handbook* to name but a couple. In addition, more precise information relating to planetary positions and so on can be found in a number of publications, a good example of which is *The Handbook of the British Astronomical Association*, as well as many of the popular astronomy magazines such as the British monthly periodicals *Sky at Night* and *Astronomy Now* and the American monthly magazines *Astronomy* and *Sky & Telescope*.

About Time

Before the late eighteenth century, the biggest problem affecting mariners sailing the seas was finding their position. Latitude was easily determined by observing the altitude of the pole star above the northern horizon. Longitude, however, was far more difficult to measure. The inability of mariners to determine their longitude often led to them getting lost, and on many occasions shipwrecked. To address this problem King Charles II established the Royal Observatory at Greenwich in 1675 and from here, Astronomers Royal began the process of measuring and cataloguing the stars as they passed due south across the Greenwich meridian.

Now mariners only needed an accurate timepiece (the chronometer invented by Yorkshire-born clockmaker John Harrison) to display GMT (Greenwich Mean Time). Working out the local standard time onboard ship and subtracting this from GMT gave the ship's longitude (west or east) from the Greenwich meridian. Therefore mariners always knew where they were at sea and the longitude problem was solved.

Astronomers use a time scale called Universal Time (UT). This is equivalent to Greenwich Mean Time and is defined by the rotation of the Earth. The Yearbook of Astronomy gives all times in UT rather than in the local time for a particular city or country. Times are expressed using the 24-hour clock, with the day beginning at midnight, denoted by 00:00. Universal Time (UT) is related to local mean time by the formula:

Local Mean Time = UT – west longitude

In practice, small differences in longitude are ignored and the observer will use local clock time which will be the appropriate Standard (or Zone) Time. As the formula indicates, places in west longitude will have a Standard Time slow on UT, while those in east longitude will have a Standard Time fast on UT. As examples we have:

Standard Time in

New Zealand	UT +12 hours
Victoria, NSW	UT +10 hours
Western Australia	UT + 8 hours
South Africa	UT + 2 hours
British Isles	UT
Newfoundland Standard Time	UT −3 hours 30 minutes
Atlantic Standard Time	UT −4 hours
Eastern Standard Time	UT −5 hours
Central Standard Time	UT −6 hours
Mountain Standard Time	UT −7 hours
Pacific Standard Time	UT −8 hours
Alaska Standard Time	UT −9 hours
Hawaii-Aleutian Standard Time	UT −10 hours

During the periods when Summer Time (also called Daylight Saving Time) is in use, one hour must be added to Standard Time to obtain the appropriate Summer/ Daylight Saving Time. For example, Pacific Daylight Time is UT −7 hours.

Using the Yearbook of Astronomy
as an Observing Guide

Notes on the Monthly Star Charts

The star charts on the following pages show the night sky throughout the year. There are two sets of charts, one for use by observers in the Northern Hemisphere and one for those in the Southern Hemisphere. The first set is drawn for latitude 52°N and can be used by observers in Europe, Canada and most of the United States. The second set is drawn for latitude 35°S and show the stars as seen from Australia and New Zealand. Twelve pairs of charts are provided for each of these latitudes.

Each pair of charts shows the entire sky as two semi-circular half-sky views, one looking north and the other looking south. A given pair of charts can be used at different times of year. For example, chart 1 shows the night sky at midnight on 21 December, but also at 2am on 21 January, 4am on 21 February and so forth. The accompanying table will enable you to select the correct chart for a given month and time of night. The caption next to each chart also lists the dates and times of night for which it is valid.

The charts are intended to help you find the more prominent constellations and other objects of interest mentioned in the monthly observing notes. To avoid the charts becoming too crowded, only stars of magnitude 4.5 or brighter are shown. This corresponds to stars that are bright enough to be seen from any dark suburban garden on a night when the Moon is not too close to full phase.

Each constellation is depicted by joining selected stars with lines to form a pattern. There is no official standard for these patterns, so you may occasionally find different patterns used in other popular astronomy books for some of the constellations.

Any map projection from a sphere onto a flat page will by necessity contain some distortions. This is true of star charts as well as maps of the Earth. The distortion on the half-sky charts is greatest near the semi-circular boundary of each chart, where it may appear to stretch constellation patterns out of shape.

The charts also show selected deep-sky objects such as galaxies, nebulae and star clusters. Many of these objects are too faint to be seen with the naked eye, and you will need binoculars or a telescope to observe them. Please refer to the table of deep-sky objects for more information.

Planetary Apparition Diagrams

The diagrams of the apparitions of Mercury and Venus show the position of the respective planet in the sky at the moment of sunrise or sunset throughout the entire apparition. Two sets of positions are plotted on each chart: for latitude 52° North (blue line) and for latitude 35° South (red line). A thin dotted line denotes the portion of the apparition which falls outside the year covered by this edition of the Yearbook. A white dot indicates the position of Venus on the first day of each month, or of Mercury on the first, eleventh and 21st of the month. The day of greatest elongation (GE) is also marked by a white dot. Note that the dots do NOT indicate the magnitude of the planet.

There are two finder charts for Mars, which comes to opposition on 8 December. The first chart, in the January notes, shows its path during the first half of 2022, when it is a morning object as it moves away from the Sun. Mars traverses 100° in ecliptic longitude during this period, moving from Ophiuchus in January to the Pisces/Aries border in late June, so the chart is based upon the ecliptic, which runs across the centre of the chart from right to left. The position of Mars is indicated on the 1st of each month from January to July. The second chart, in the December notes, uses a conventional R.A./Dec. grid and follows Mars from 1 July 2022 to 1 February 2023 as it crosses Aries and performs its retrograde loop among the stars of Taurus. On this chart, the position of Mars is indicated on the 1st of each month, and also at its stationary points and on the date of opposition. Stars are shown to magnitude 5.5 on both charts. Note that the sizes of the Mars dots do NOT indicate its magnitude.

The finder charts for Jupiter, Saturn, Uranus and Neptune show the paths of the planets throughout the year. The position of each planet is indicated at opposition and at stationary points, as well as the start and end of the year and on the 1st of each month (1st of April, July and October only for Uranus and Neptune) where these dates do not fall too close to an event that is already marked. Stars are shown to magnitude 5.5 on the charts for Jupiter and Saturn. On the Uranus chart, stars are shown to magnitude 8; on the Neptune chart, the limiting magnitude is 10. In both cases, this is approximately two magnitudes fainter than the planet itself. Right Ascension and Declination scales are shown for the epoch J2000 to allow comparison with modern star charts. Note that the sizes of the dots denoting the planets do NOT indicate their magnitudes.

Selecting the Correct Charts

The table below shows which of the charts to use for particular dates and times throughout the year and will help you to select the correct pair of half-sky charts for any combination of month and time of night.

The Earth takes 23 hours 56 minutes (and 4 seconds) to rotate once around its axis with respect to the fixed stars. Because this is around four minutes shorter than a full 24 hours, the stars appear to rise and set about 4 minutes earlier on each successive day, or around an hour earlier each fortnight. Therefore, as well as showing the stars at 10pm (22h in 24-hour notation) on 21 January, chart 1 also depicts the sky at 9pm (21h) on 6 February, 8pm (20h) on 21 February and 7pm (19h) on 6 March.

The times listed do not include summer time (daylight saving time), so if summer time is in force you must subtract one hour to obtain standard time (GMT if you are in the United Kingdom) before referring to the chart. For example, to find the correct chart for mid-September in the northern hemisphere at 3am summer time, first of all subtract one hour to obtain 2am (2h) standard time. Then you can consult the table, where you will find that you should use chart 11.

The table does not indicate sunrise, sunset or twilight. In northern temperate latitudes, the sky is still light at 18h and 6h from April to September, and still light at 20h and 4h from May to August. In Australia and New Zealand, the sky is still light at 18h and 6h from October to March, and in twilight (with only bright stars visible) at 20h and 04h from November to January.

Local Time	18h	20h	22h	0h	2h	4h	6h
January	11	12	1	2	3	4	5
February	12	1	2	3	4	5	6
March	1	2	3	4	5	6	7
April	2	3	4	5	6	7	8
May	3	4	5	6	7	8	9
June	4	5	6	7	8	9	10
July	5	6	7	8	9	10	11
August	6	7	8	9	10	11	12
September	7	8	9	10	11	12	1
October	8	9	10	11	12	1	2
November	9	10	11	12	1	2	3
December	10	11	12	1	2	3	4

Legend to the Star Charts

STARS		DEEP-SKY OBJECTS	
Symbol	Magnitude	Symbol	Type of object
•	0 or brighter	⁕	Open star cluster
•	1	○	Globular star cluster
•	2	□	Nebula
•	3	▦	Cluster with nebula
•	4	⊙	Planetary nebula
·	5	⸝	Galaxy
✦	Double star		Magellanic Clouds
◉	Variable star		

Star Names

There are over 200 stars with proper names, most of which are of Roman, Greek or Arabic origin although only a couple of dozen or so of these names are used regularly. Examples include Arcturus in Boötes, Castor and Pollux in Gemini and Rigel in Orion.

A system whereby Greek letters were assigned to stars was introduced by the German astronomer and celestial cartographer Johann Bayer in his star atlas Uranometria, published in 1603. Bayer's system is applied to the brighter stars within any particular constellation, which are given a letter from the Greek alphabet followed by the genitive case of the constellation in which the star is located. This genitive case is simply the Latin form meaning 'of' the constellation. Examples are the stars Alpha Boötis and Beta Centauri which translate literally as 'Alpha of Boötes' and 'Beta of the Centaur'.

As a general rule, the brightest star in a constellation is labelled Alpha (α), the second brightest Beta (β), and the third brightest Gamma (γ) and so on, although there are some constellations where the system falls down. An example is Gemini where the principal star (Pollux) is designated Beta Geminorum, the second brightest (Castor) being known as Alpha Geminorum.

There are only 24 letters in the Greek alphabet, the consequence of which was that the fainter naked eye stars needed an alternative system of classification. The system in popular use is that devised by the first Astronomer Royal John Flamsteed in which the stars in each constellation are listed numerically in order from west to east. Although many of the brighter stars within any particular constellation will have both Greek letters and Flamsteed numbers, the latter are generally used only when a star does not have a Greek letter.

The Greek Alphabet

α	Alpha	ι	Iota	ρ	Rho
β	Beta	κ	Kappa	σ	Sigma
γ	Gamma	λ	Lambda	τ	Tau
δ	Delta	μ	Mu	υ	Upsilon
ε	Epsilon	ν	Nu	φ	Phi
ζ	Zeta	ξ	Xi	χ	Chi
η	Eta	ο	Omicron	ψ	Psi
θ	Theta	π	Pi	ω	Omega

The Names of the Constellations

On clear, dark, moonless nights, the sky seems to teem with stars although in reality you can never see more than a couple of thousand or so at any one time when looking with the unaided eye. Each and every one of these stars belongs to a particular constellation, although the constellations that we see in the sky, and which grace the pages of star atlases, are nothing more than chance alignments. The stars that make up the constellations are often situated at vastly differing distances from us and only appear close to each other, and form the patterns that we see, because they lie in more or less the same direction as each other as seen from Earth.

A large number of the constellations are named after mythological characters, and were given their names thousands of years ago. However, those star groups lying close to the south celestial pole were discovered by Europeans only during the last few centuries, many of these by explorers and astronomers who mapped the stars during their journeys to lands under southern skies. This resulted in many of the newer constellations having modern-sounding names, such as Octans (the Octant) and Microscopium (the Microscope), both of which were devised by the French astronomer Nicolas Louis De La Caille during the early 1750s.

Over the centuries, many different suggestions for new constellations have been put forward by astronomers who, for one reason or another, felt the need to add new groupings to star charts and to fill gaps between the traditional constellations. Astronomers drew up their own charts of the sky, incorporating their new groups into them. A number of these new constellations had cumbersome names, notable examples including Officina Typographica (the Printing Shop) introduced by the German astronomer Johann Bode in 1801; Sceptrum Brandenburgicum (the Sceptre of Brandenburg) introduced by the German astronomer Gottfried Kirch in 1688; Taurus Poniatovii (Poniatowski's Bull) introduced by the Polish-Lithuanian astronomer Martin Odlanicky Poczobut in 1777; and Quadrans Muralis (the Mural Quadrant) devised by the French astronomer Joseph-Jerôme de Lalande in1795. Although these have long since been rejected, the latter has been immortalised by the annual Quadrantid meteor shower, the radiant of which lies in an area of sky formerly occupied by Quadrans Muralis.

During the 1920s the International Astronomical Union (IAU) systemised matters by adopting an official list of 88 accepted constellations, each with official spellings and abbreviations. Precise boundaries for each constellation were then drawn up so that every point in the sky belonged to a particular constellation.

The abbreviations devised by the IAU each have three letters which in the majority of cases are the first three letters of the constellation name, such as AND for Andromeda, EQU for Equuleus, HER for Hercules, ORI for Orion and so on. This trend is not strictly adhered to in cases where confusion may arise. This happens with the two constellations Leo (abbreviated LEO) and Leo Minor (abbreviated LMI). Similarly, because Triangulum (TRI) may be mistaken for Triangulum Australe, the latter is abbreviated TRA. Other instances occur with Sagitta (SGE) and Sagittarius (SGR) and with Canis Major (CMA) and Canis Minor (CMI) where the first two letters from the second names of the constellations are used. This is also the case with Corona Australis (CRA) and Corona Borealis (CRB) where the first letter of the second name of each constellation is incorporated. Finally, mention must be made of Crater (CRT) which has been abbreviated in such a way as to avoid confusion with the aforementioned CRA (Corona Australis)..

The table shown on the following pages contains the name of each of the 88 constellations together with the translation and abbreviation of the constellation name. The constellations depicted on the monthly star charts are identified with their abbreviations rather than the full constellation names.

The Constellations

Andromeda	Andromeda	AND	Delphinus	The Dolphin	DEL
Antlia	The Air Pump	ANT	Dorado	The Goldfish	DOR
Apus	The Bird of Paradise	APS	Draco	The Dragon	DRA
			Equuleus	The Foal	EQU
Aquarius	The Water Carrier	AQR	Eridanus	The River	ERI
Aquila	The Eagle	AQL	Fornax	The Furnace	FOR
Ara	The Altar	ARA	Gemini	The Twins	GEM
Aries	The Ram	ARI	Grus	The Crane	GRU
Auriga	The Charioteer	AUR	Hercules	Hercules	HER
Boötes	The Herdsman	BOO	Horologium	The Pendulum Clock	HOR
Caelum	The Graving Tool	CAE			
Camelopardalis	The Giraffe	CAM	Hydra	The Water Snake	HYA
Cancer	The Crab	CNC	Hydrus	The Lesser Water Snake	HYI
Canes Venatici	The Hunting Dogs	CVN	Indus	The Indian	IND
Canis Major	The Great Dog	CMA			
Canis Minor	The Little Dog	CMI	Lacerta	The Lizard	LAC
Capricornus	The Goat	CAP	Leo	The Lion	LEO
Carina	The Keel	CAR	Leo Minor	The Lesser Lion	LMI
Cassiopeia	Cassiopeia	CAS	Lepus	The Hare	LEP
Centaurus	The Centaur	CEN	Libra	The Scales	LIB
Cepheus	Cepheus	CEP	Lupus	The Wolf	LUP
Cetus	The Whale	CET	Lynx	The Lynx	LYN
Chamaeleon	The Chameleon	CHA	Lyra	The Lyre	LYR
Circinus	The Pair of Compasses	CIR	Mensa	The Table Mountain	MEN
			Microscopium	The Microscope	MIC
Columba	The Dove	COL	Monoceros	The Unicorn	MON
Coma Berenices	Berenice's Hair	COM	Musca	The Fly	MUS
Corona Australis	The Southern Crown	CRA	Norma	The Level	NOR
Corona Borealis	The Northern Crown	CRB	Octans	The Octant	OCT
			Ophiuchus	The Serpent Bearer	OPH
Corvus	The Crow	CRV	Orion	Orion	ORI
Crater	The Cup	CRT	Pavo	The Peacock	PAV
Crux	The Cross	CRU	Pegasus	Pegasus	PEG
Cygnus	The Swan	CYG	Perseus	Perseus	PER

Phoenix	The Phoenix	PHE	Sextans	The Sextant	SEX
Pictor	The Painter's Easel	PIC	Taurus	The Bull	TAU
Pisces	The Fish	PSC	Telescopium	The Telescope	TEL
Piscis Austrinus	The Southern Fish	PSA	Triangulum	The Triangle	TRI
Puppis	The Stern	PUP	Triangulum Australe	The Southern Triangle	TRA
Pyxis	The Mariner's Compass	PYX	Tucana	The Toucan	TUC
Reticulum	The Net	RET	Ursa Major	The Great Bear	UMA
Sagitta	The Arrow	SGE	Ursa Minor	The Little Bear	UMI
Sagittarius	The Archer	SGR	Vela	The Sail	VEL
Scorpius	The Scorpion	SCO	Virgo	The Virgin	VIR
Sculptor	The Sculptor	SCL	Volans	The Flying Fish	VOL
Scutum	The Shield	SCT	Vulpecula	The Fox	VUL
Serpens Caput and Cauda	The Serpent	SER			

The Monthly Star Charts

Northern Hemisphere Star Charts

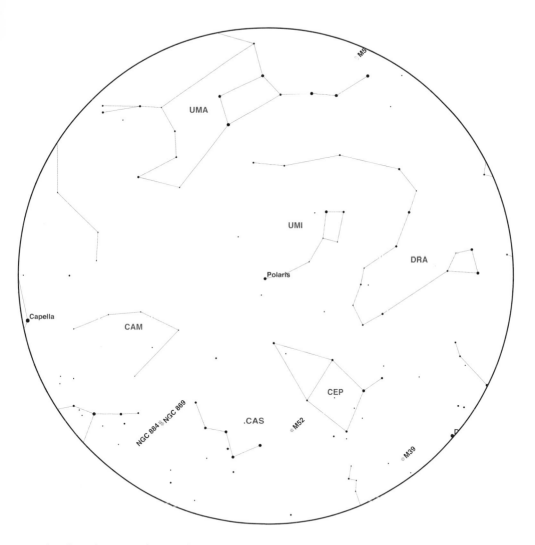

This chart shows stars lying at declinations between +45 and +90 degrees. These constellations are circumpolar for observers in Europe and North America.

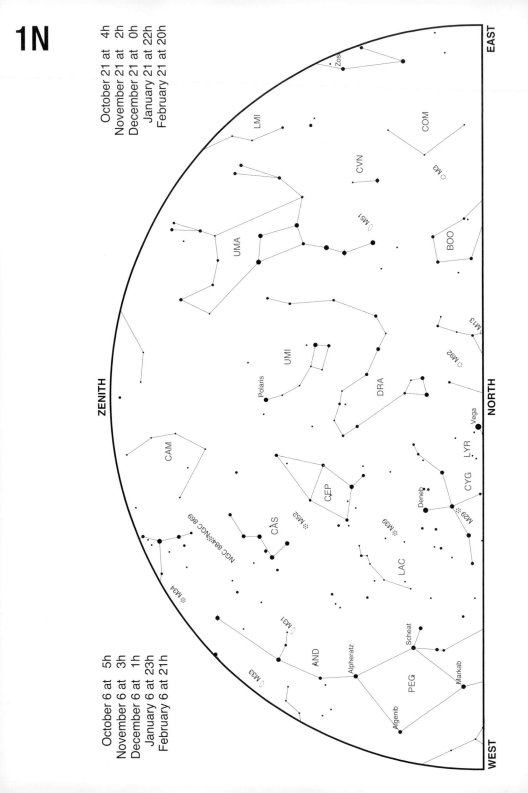

1N

October 21 at 4h
November 21 at 2h
December 21 at 0h
January 21 at 22h
February 21 at 20h

October 6 at 5h
November 6 at 3h
December 6 at 1h
January 6 at 23h
February 6 at 21h

EAST

ZENITH

NORTH

WEST

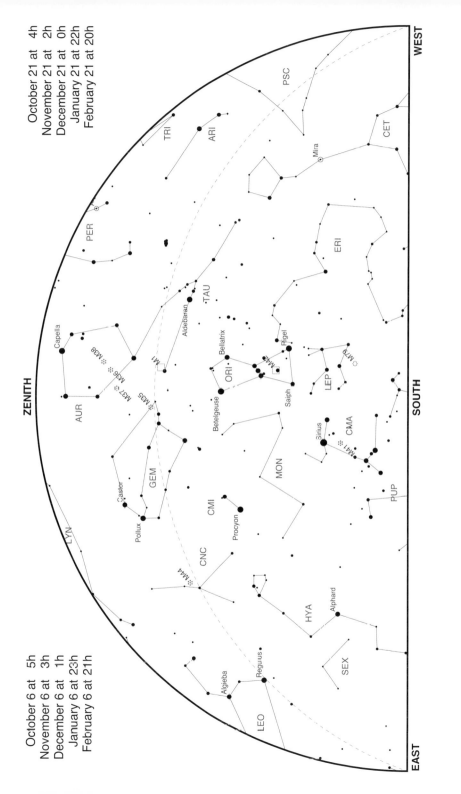

1S

WEST

PSC

CET

Mira

TRI

ARI

PER

ERI

TAU

Aldebaran

Capella

ZENITH

AUR

M38
M36
M37

M35

M1

Bellatrix

Rigel

M42

ORI

Betelgeuse

Saiph

LEP

M79

Castor

GEM

Pollux

LYN

MON

CMI

Procyon

Sirius

CMA

M41

PUP

SOUTH

CNC

M44

HYA

Alphard

SEX

Regulus

Algieba

LEO

EAST

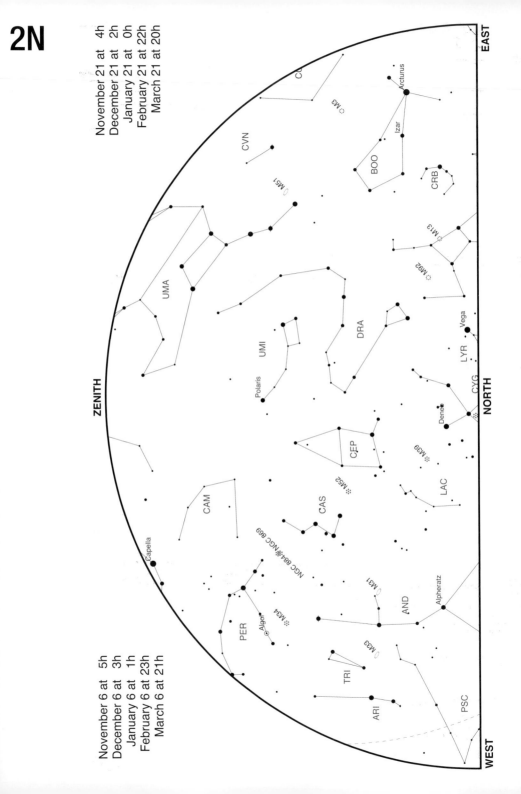

2N

November 21 at 4h
December 21 at 2h
January 21 at 0h
February 21 at 22h
March 21 at 20h

November 6 at 5h
December 6 at 3h
January 6 at 1h
February 6 at 23h
March 6 at 21h

EAST
ZENITH
NORTH
WEST

Arcturus
Izar
BOO
CRB
M3
CO
CVN
M51
UMA
M13
M92
DRA
UMI
Polaris
Vega
LYR
CYG
Deneb
CEP
M39
CAM
LAC
Capella
CAS
M52
NGC 884 · NGC 869
PER
Algol
M34
TRI
M31
AND
Alpheratz
M33
ARI
PSC

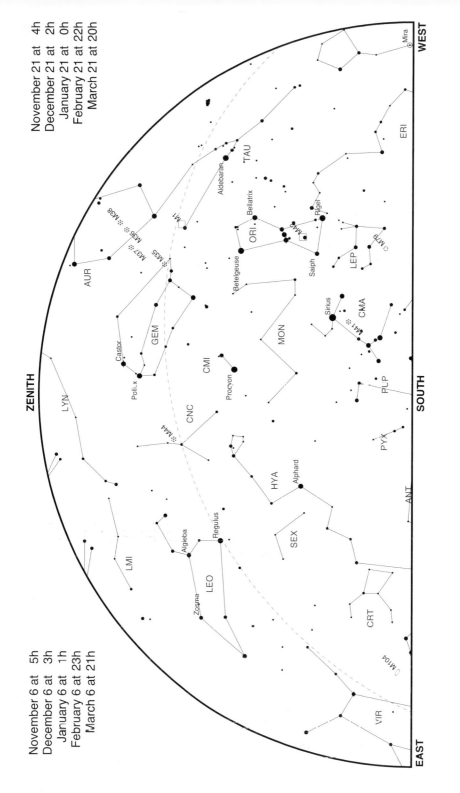

2S

WEST

EAST

SOUTH

ZENITH

Mira

TAU
Aldebaran
Bellatrix
ORI
Rigel
M42
Betelgeuse
Saiph
LEP
α LEP
M79
ERI
Sirius
CMA
M41
AUR
M38
M36
M37
M35
M1
GEM
Castor
Poll.x
CNC
M44
CMI
Procyon
MON
PUP
PYX
ANT
LYN
LMI
LEO
Algieba
Regulus
Zosma
SEX
HYA
Alphard
CRT
M104
VIR

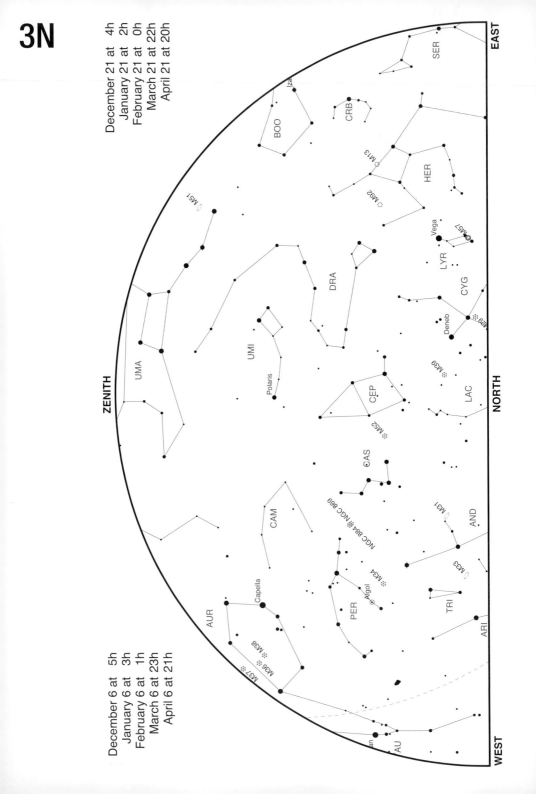

3N

December 21 at 4h
January 21 at 2h
February 21 at 0h
March 21 at 22h
April 21 at 20h

December 6 at 5h
January 6 at 3h
February 6 at 1h
March 6 at 23h
April 6 at 21h

ZENITH

EAST

NORTH

WEST

SER

CRB

BOO

Izar

HER

M13

M92

Vega

LYR

M57

CYG

Deneb

M29

M39

DRA

M51

UMI

Polaris

UMA

CEP

M52

CAS

LAC

CAM

M31

NGC 884 & NGC 869

AND

M33

PER

Algol

M34

TRI

Capella

AUR

ARI

M38

M36

M37

AU

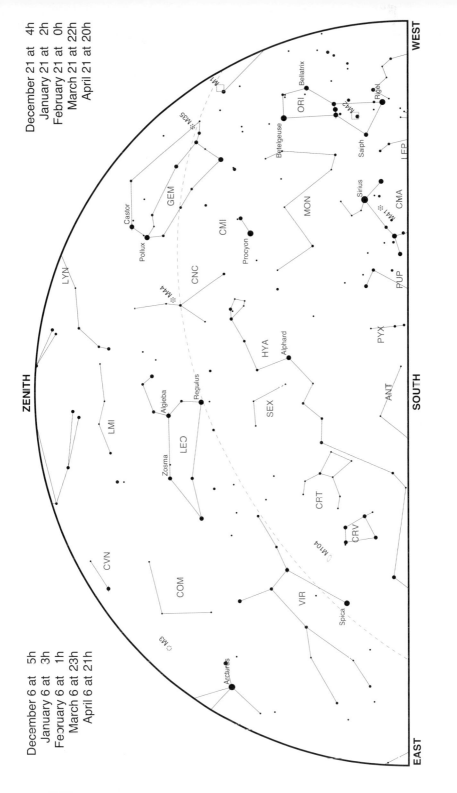

3S

WEST

December 21 at 4h
January 21 at 2h
February 21 at 0h
March 21 at 22h
April 21 at 20h

ZENITH

December 6 at 5h
January 6 at 3h
February 6 at 1h
March 6 at 23h
April 6 at 21h

EAST

SOUTH

4N

January 21 at 4h
February 21 at 2h
March 21 at 0h
April 21 at 22h
May 21 at 20h

January 6 at 5h
February 6 at 3h
March 6 at 1h
April 6 at 23h
May 6 at 21h

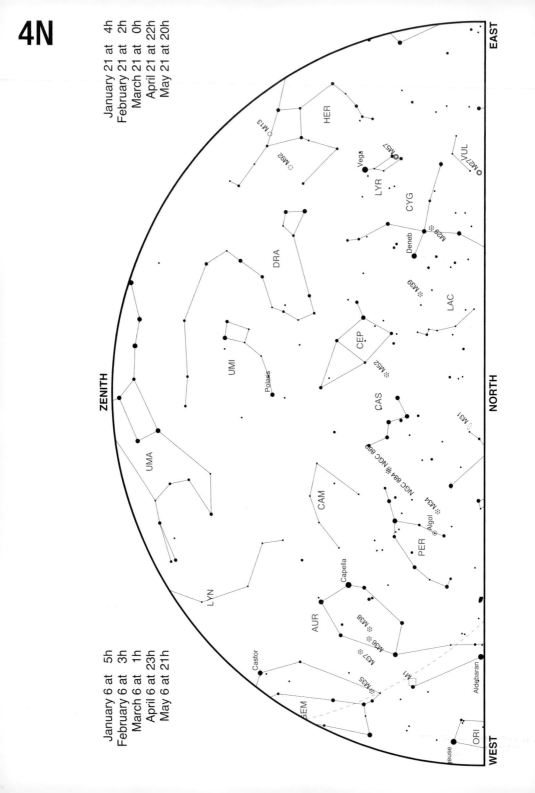

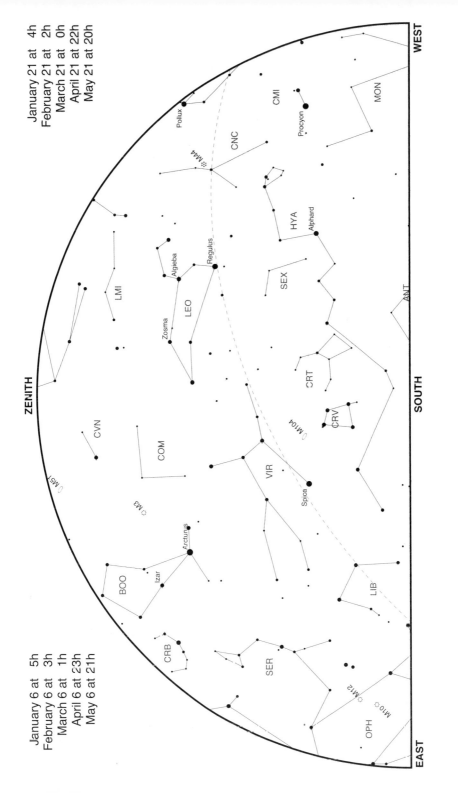

4S

January 21 at 4h
February 21 at 2h
March 21 at 0h
April 21 at 22h
May 21 at 20h

January 6 at 5h
February 6 at 3h
March 6 at 1h
April 6 at 23h
May 6 at 21h

WEST

EAST

SOUTH

ZENITH

Pollux
CMI
MON
Procyon
CNC
M44
HYA
Alphard
Algieba
Regulus
SEX
LMI
LEO
Zosma
ANT
CRT
CVN
M51
COM
CRV
M104
VIR
Spica
M3
BOO
Arcturus
Izar
LIB
CRB
SER
M12
M10
OPH

5N

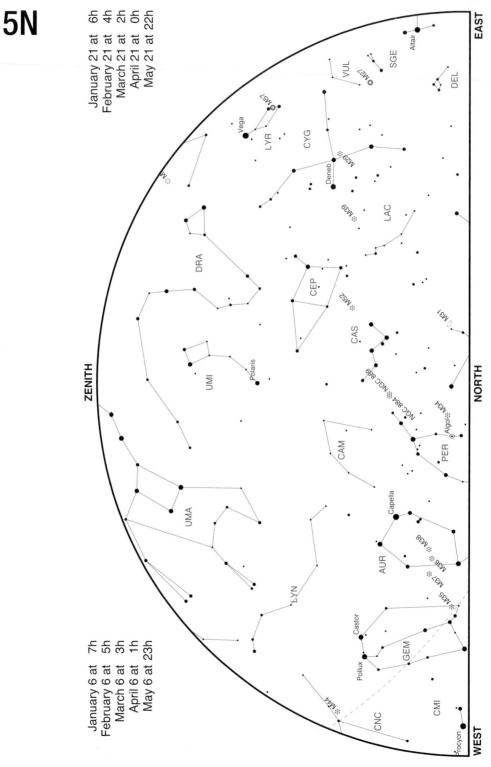

January 21 at 6h
February 21 at 4h
March 21 at 2h
April 21 at 0h
May 21 at 22h

January 6 at 7h
February 6 at 5h
March 6 at 3h
April 6 at 1h
May 6 at 23h

EAST

Altair

SGE

VUL
M27

DEL

Vega
M57
LYR
CYG

M29
Deneb

M39
LAC

CEP

M52
CAS

M31

DRA

Polaris
UMI

ZENITH

NGC 884 NGC 869
M34
Algol
PER
CAM

NORTH

Capella

UMA

AUR
M36 M38

M37

LYN
M35

Castor

Pollux
GEM

CNC
M44

CMI

Procyon

WEST

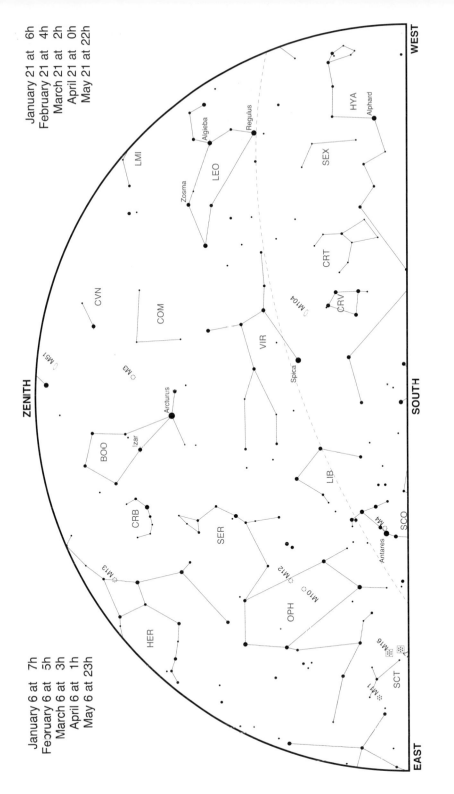

5S

WEST

EAST

SOUTH

ZENITH

LMI

LEO

Algieba

Regulus

Zosma

SEX

HYA

Alphard

CVN

COM

CRT

M104

CRV

VIR

Spica

M51

M3

Arcturus

Izar

BOO

LIB

CRB

SER

M4

SCO

Antares

M13

HER

M12

M10

OPH

M16

M11

SCT

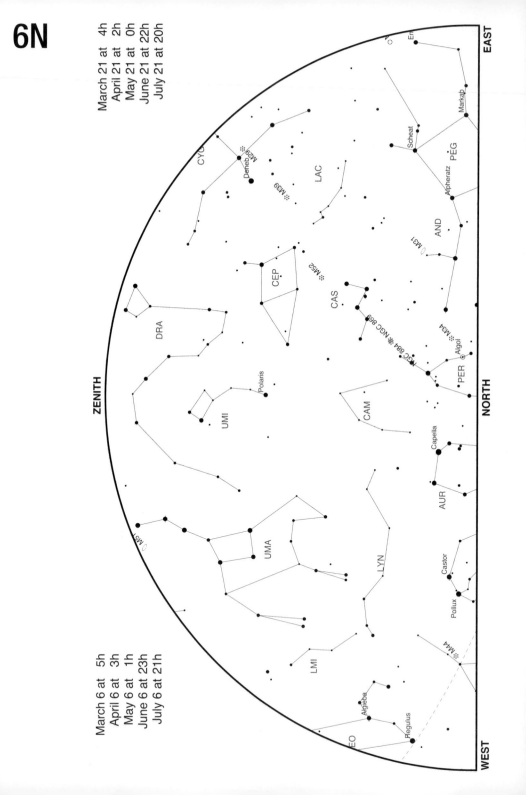

6N

EAST

ZENITH

NORTH

WEST

March 6 at 5h
April 6 at 3h
May 6 at 1h
June 6 at 23h
July 6 at 21h

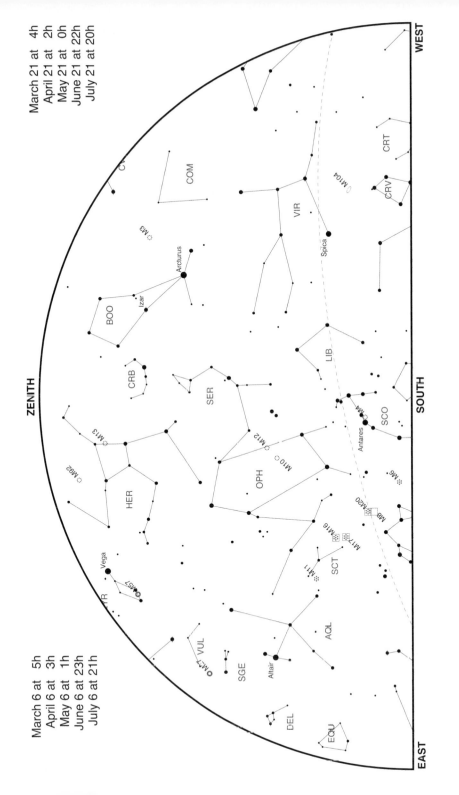

6S

WEST

March 21 at 4h
April 21 at 2h
May 21 at 0h
June 21 at 22h
July 21 at 20h

ZENITH

March 6 at 5h
April 6 at 3h
May 6 at 1h
June 6 at 23h
July 6 at 21h

EAST

SOUTH

COM
CRT
CRV
M104
VIR
Spica
Arcturus
Izar
BOO
M3
LIB
CRB
SER
SCO
M4
Antares
M12
M10
OPH
M6
HER
M92
M13
M8 M20
M17 M16
SCT
M11
Vega
M57
LYR
VUL
SGE
Altair
AQL
DEL
EQU

7N

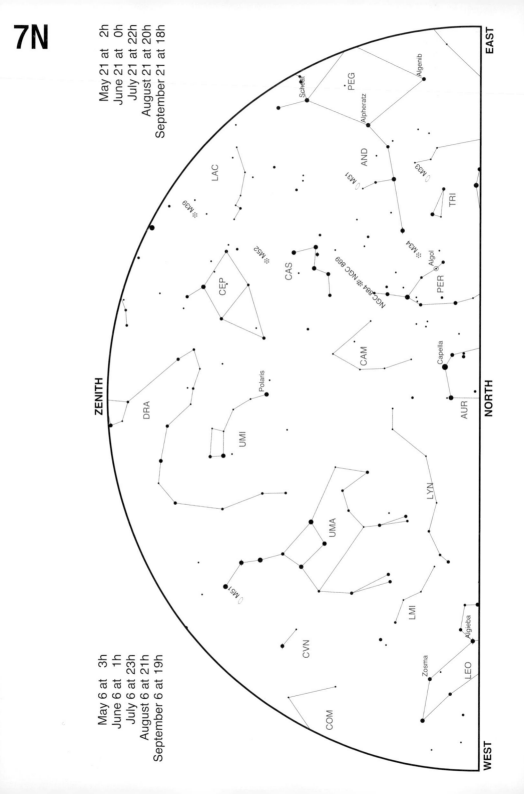

May 21 at 2h
June 21 at 0h
July 21 at 22h
August 21 at 20h
September 21 at 18h

May 6 at 3h
June 6 at 1h
July 6 at 23h
August 6 at 21h
September 6 at 19h

EAST

WEST

NORTH

ZENITH

PEG
Scheat
Algenib
Alpheratz
AND
M31
M33
TRI
LAC
M39
M52
CAS
M34
Algol
NGC 884 NGC 869
PER
CEP
CAM
Capella
AUR
DRA
Polaris
UMI
LYN
UMA
M51
LMI
CVN
COM
Zosma
Algieba
LEO

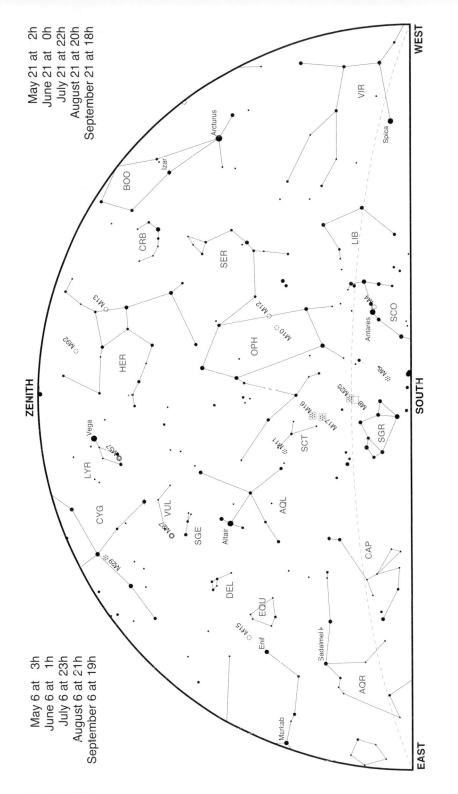

7S

May 21 at 2h
June 21 at 0h
July 21 at 22h
August 21 at 20h
September 21 at 18h

May 6 at 3h
June 6 at 1h
July 6 at 23h
August 6 at 21h
September 6 at 19h

ZENITH

EAST

SOUTH

VIR
Spica
BOO
Arcturus
Izar
CRB
SER
LIB
M13
M92
HER
OPH
M10
M12
M4
SCO
Antares
M6
M25
M8
M22
SGR
M17
M16
M57
Vega
LYR
M11
SCT
CYG
M27
VUL
M29
SGE
Altair
AQL
DEL
EQU
M15
Enif
Sadalmelik
CAP
Merkab
AQR

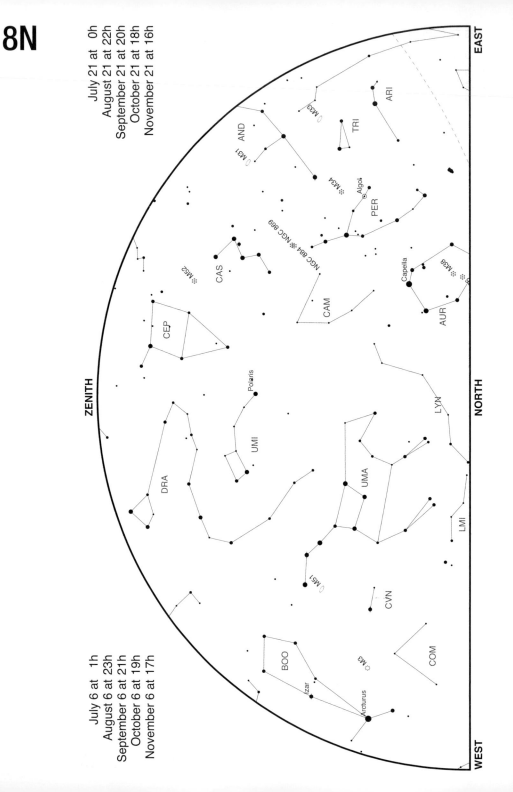

8N

July 6 at 1h
August 6 at 23h
September 6 at 21h
October 6 at 19h
November 6 at 17h

ZENITH

EAST

NORTH

WEST

AND
M31
TRI
M33
ARI
M34
Algol
PER
NGC 884 NGC 869
CAS
M52
CAM
Capella
M38
M36
AUR
CEP
Polaris
UMI
LYN
DRA
UMA
LMI
M51
CVN
M3
BOO
Izar
Arcturus
COM

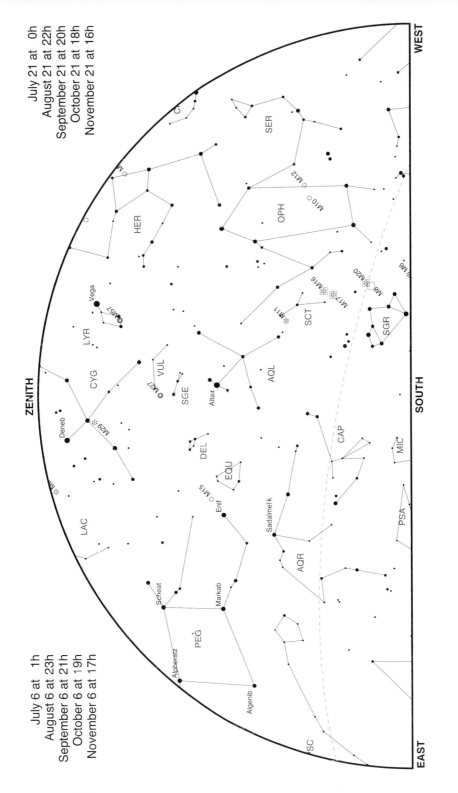

8S

WEST

ZENITH

EAST

SOUTH

July 21 at 0h
August 21 at 22h
September 21 at 20h
October 21 at 18h
November 21 at 16h

July 6 at 1h
August 6 at 23h
September 6 at 21h
October 6 at 19h
November 6 at 17h

SER
OPH
HER
LYR
Vega
M57
CYG
VUL
M27
SGE
Altair
AQL
SCT
M11
M17
M16
M20
M8
M6
SGR
Deneb
M29
LAC
DEL
EQU
M15
Enif
CAP
MIC
Scheat
Markab
PEG
Alpheratz
Algenib
AQR
Sadalmelik
PSA
M10
M12
SC

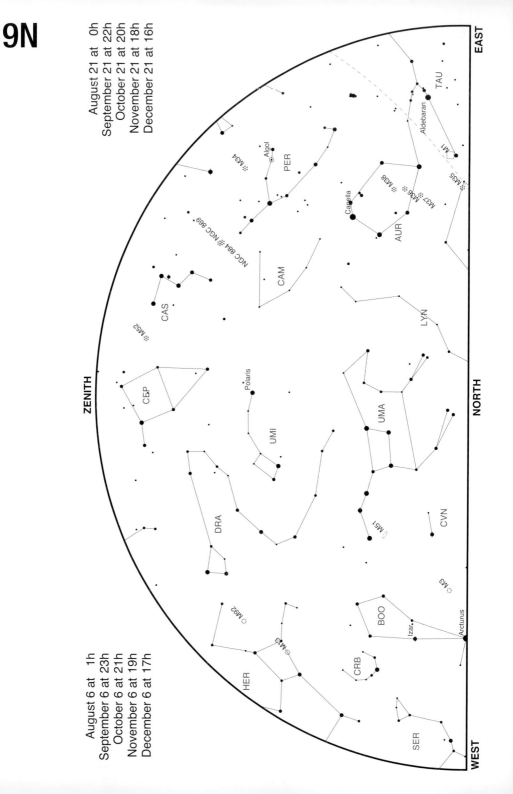

9N

August 21 at 0h
September 21 at 22h
October 21 at 20h
November 21 at 18h
December 21 at 16h

August 6 at 1h
September 6 at 23h
October 6 at 21h
November 6 at 19h
December 6 at 17h

EAST

ZENITH

NORTH

WEST

TAU
Aldebaran
M1
M35
Algol
M34
PER
M37
M36
M38
Capella
AUR
M52
CAM
NGC 884 NGC 869
CAS
LYN
CEP
Polaris
UMI
UMA
DRA
M51
CVN
M3
M92
M13
Arcturus
Izar
BOO
HER
CRB
SER

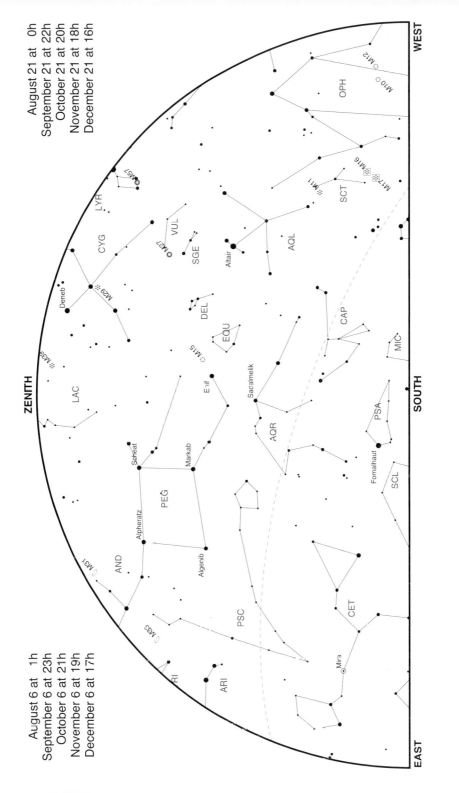

9S

August 21 at 0h
September 21 at 22h
October 21 at 20h
November 21 at 18h
December 21 at 16h

August 6 at 1h
September 6 at 23h
October 6 at 21h
November 6 at 19h
December 6 at 17h

WEST

ZENITH

EAST

SOUTH

OPH
M12
M10
SCT
M16
M17
M11
AQL
Altair
LYR
M57
CYG
Deneb
M29
VUL
M27
SGE
DEL
EQU
M15
CAP
MIC
Enif
Sadalmelik
AQR
PSA
Fomalhaut
SCL
LAC
M39
Scheat
Markab
PEG
Alpheratz
AND
M31
Algenib
PSC
ARI
M33
ARI
CET
Mira

10N

August 21 at 2h
September 21 at 0h
October 21 at 22h
November 21 at 20h
December 21 at 18h

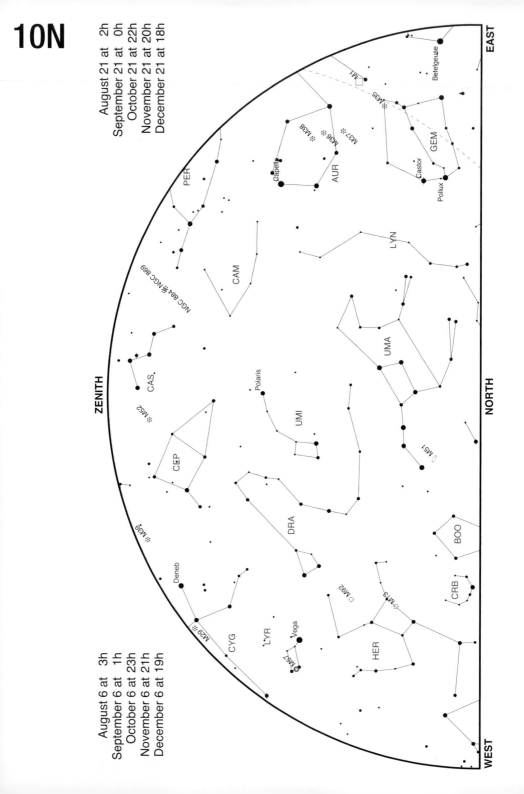

EAST

ZENITH

NORTH

WEST

August 6 at 3h
September 6 at 1h
October 6 at 23h
November 6 at 21h
December 6 at 19h

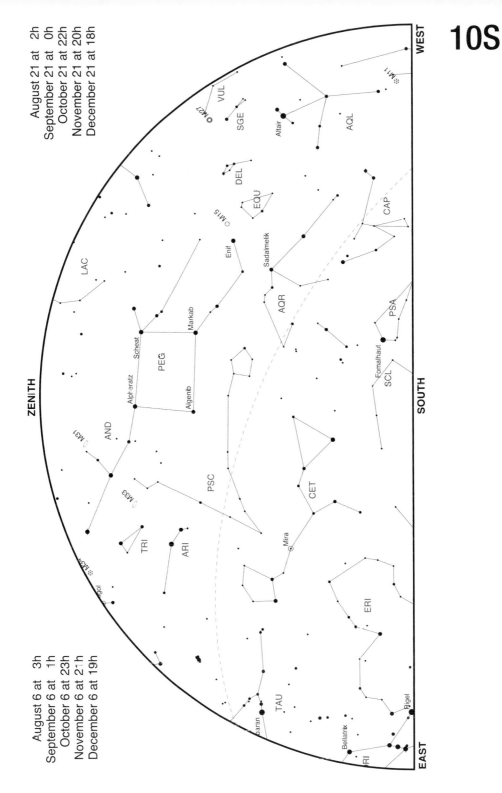

10S

August 21 at 2h
September 21 at 0h
October 21 at 22h
November 21 at 20h
December 21 at 18h

August 6 at 3h
September 6 at 1h
October 6 at 23h
November 6 at 21h
December 6 at 19h

WEST

ZENITH

EAST

SOUTH

VUL
M27
SGE
Altair
AQL
M11

DEL
EQU
CAP

LAC
M15
Enif
Sadalmelik
AQR
PSA

Scheat
Markab
PEG
Alpheratz
Algenib
Fomalhaut
SCL

AND
M31
M33
PSC
CET
Mira

TRI
ARI
M34
Algol

ERI
TAU
Aldebaran
Bellatrix
Rigel
ORI

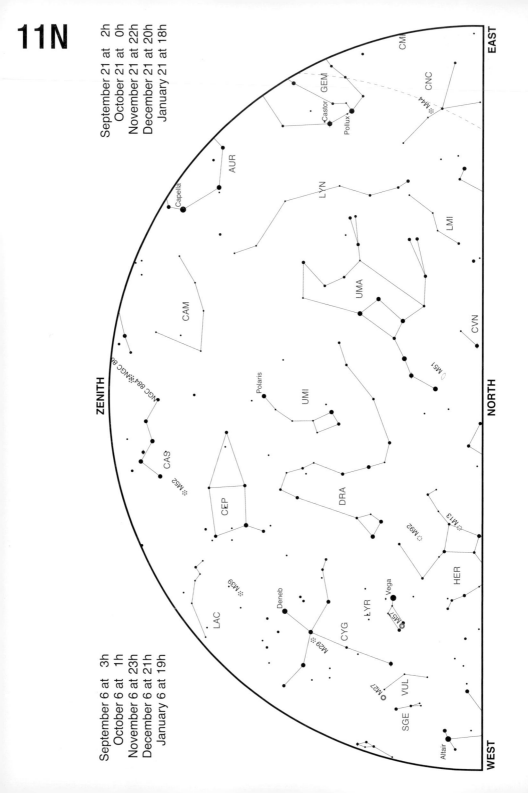

11N

September 21 at 2h
October 21 at 0h
November 21 at 22h
December 21 at 20h
January 21 at 18h

September 6 at 3h
October 6 at 1h
November 6 at 23h
December 6 at 21h
January 6 at 19h

EAST

WEST

NORTH

ZENITH

CMI
CNC
M44
GEM
Castor
Pollux
AUR
Capella
LYN
LMI
CAM
UMA
CVN
M51
NGC 884 NGC 869
Polaris
UMI
CAS
M52
CEP
DRA
M92
M13
HER
LAC
M39
Deneb
CYG
M29
LYR
Vega
M57
VUL
SGE
M27
Altair

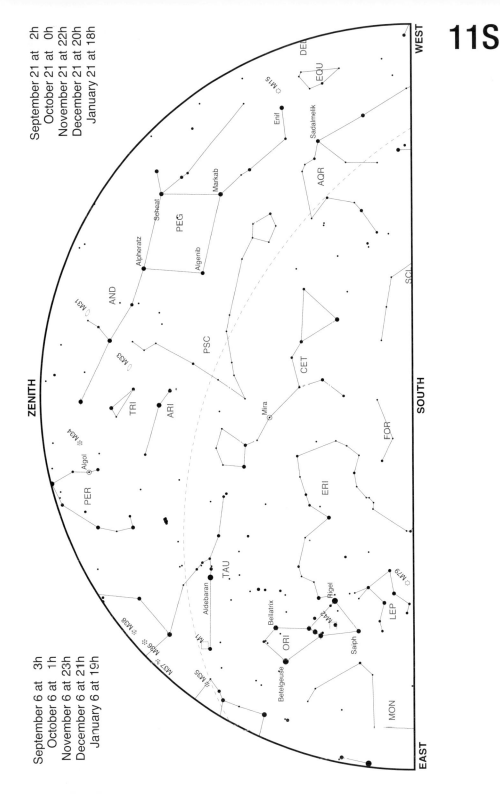

11S

September 21 at 2h
October 21 at 0h
November 21 at 22h
December 21 at 20h
January 21 at 18h

September 6 at 3h
October 6 at 1h
November 6 at 23h
December 6 at 21h
January 6 at 19h

WEST

ZENITH

SOUTH

EAST

DEL
EQU
M15
Enif
Sadalmelik
Markab
Scheat
PEG
AQR
Algenib
Alpheratz
AND
M31
PSC
M33
SCL
TRI
ARI
CET
Mira
FOR
M34
Algol
PER
ERI
M38
M36
M37
TAU
M1
Aldebaran
M35
Bellatrix
M42
ORI
Rigel
LEP
M79
Saiph
Betelgeuse
MON

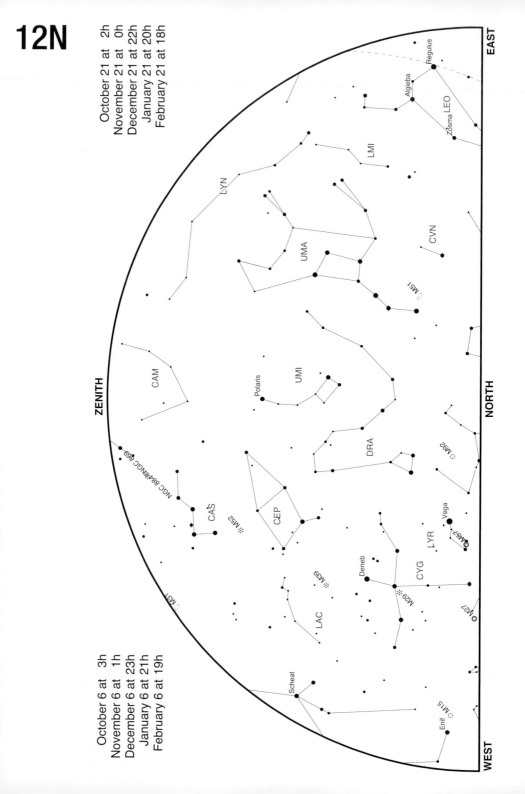

12N

October 6 at 3h
November 6 at 1h
December 6 at 23h
January 6 at 21h
February 6 at 19h

ZENITH

EAST

NORTH

WEST

LEO
Regulus
Algieba
Zosma
LMI
LYN
CVN
M51
UMA
CAM
Polaris
UMI
DRA
M92
NGC 884 NGC 869
CAS
M52
CEP
Vega
LYR
M57
M39
Deneb
CYG
M29
LAC
M31
M15
Scheat
Enif
M27

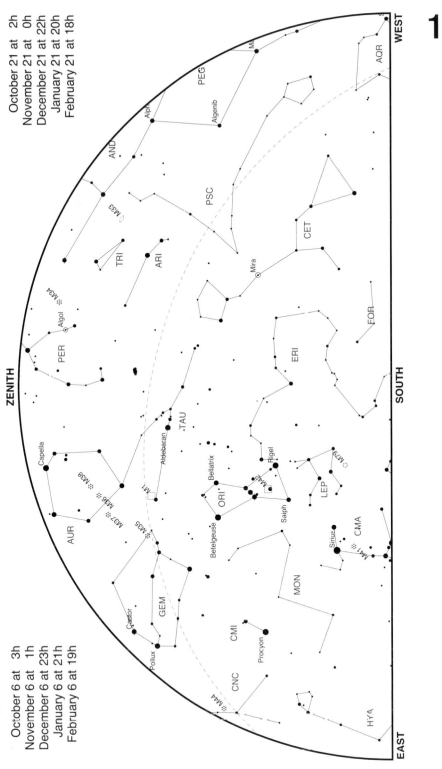

12S

October 21 at 2h
November 21 at 0h
December 21 at 22h
January 21 at 20h
February 21 at 18h

October 6 at 3h
November 6 at 1h
December 6 at 23h
January 6 at 21h
February 6 at 19h

WEST

EAST

SOUTH

ZENITH

AQR

PEG

Algenib

AND

Alph

PSC

CET

Mira

M33

TRI

ARI

M34

Algol

PER

Capella

AUR

M38

M37

M36

M35

M1

TAU

Aldebaran

Bellatrix

ORI

Betelgeuse

Rigel

M42

Saiph

ERI

FOR

LEP

M79

CMA

Sirius

M41

MON

CMI

Procyon

GEM

Castor

Pollux

CNC

M44

HYA

Southern Hemisphere Star Charts

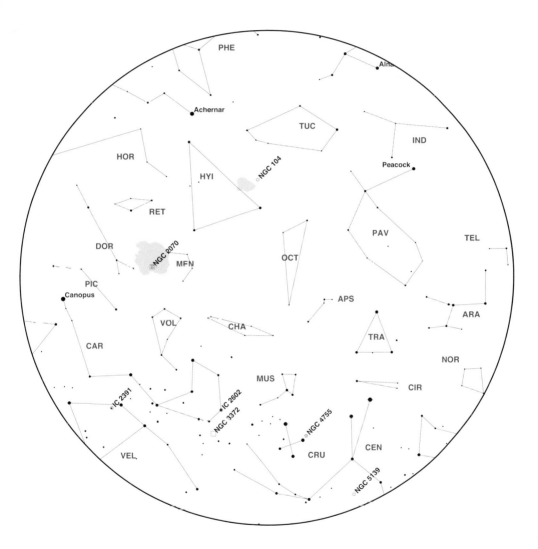

This chart shows stars lying at declinations between −45 and −90 degrees. These constellations are circumpolar for observers in Australia and New Zealand.

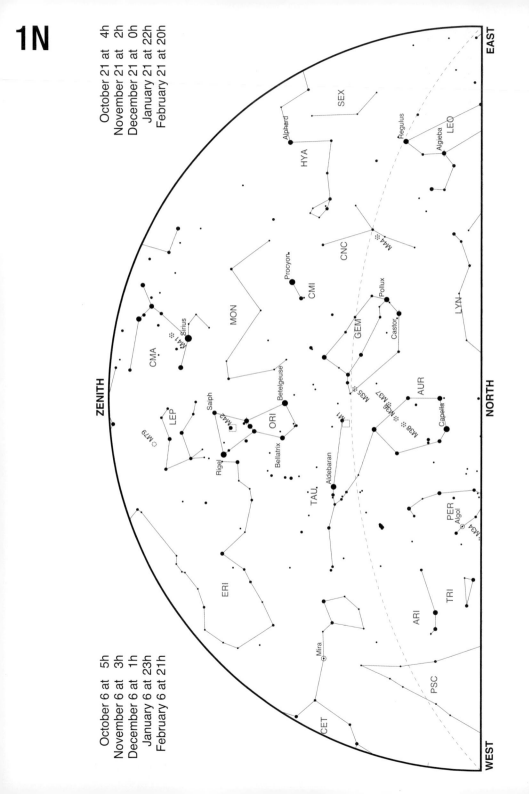

October 21 at 4h
November 21 at 2h
December 21 at 0h
January 21 at 22h
February 21 at 20h

October 6 at 5h
November 6 at 3h
December 6 at 1h
January 6 at 23h
February 6 at 21h

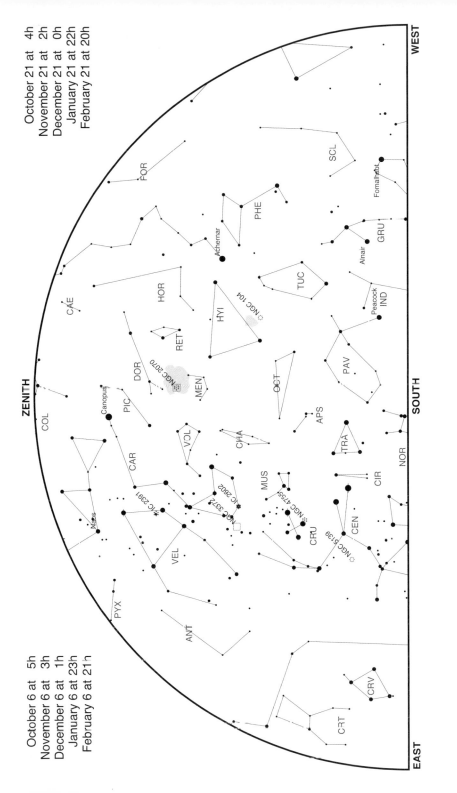

1S

WEST

ZENITH

SOUTH

EAST

FOR
SCL
PHE
Achernar
Fomalhaut
HYI
NGC 104
TUC
GRU
Alnair
HOR
CAE
RET
MEN
DOR
NGC 2070
OCT
PAV
IND
Peacock
APS
PIC
Canopus
CHA
COL
VOL
TRA
CAR
MUS
CIR
NGC 4755
CRU
NGC 5139
CEN
NOR
Mars
IC 2391
NGC 3372
IC 2602
VEL
PYX
ANT
CRV
CRT

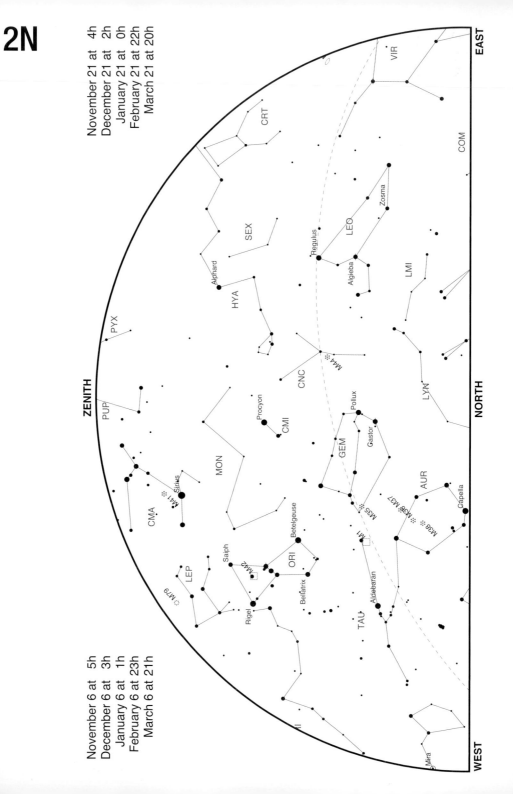

2N

November 21 at 4h
December 21 at 2h
January 21 at 0h
February 21 at 22h
March 21 at 20h

November 6 at 5h
December 6 at 3h
January 6 at 1h
February 6 at 23h
March 6 at 21h

EAST

VIR

CRT

COM

SEX

LEO

Zosma

Regulus

Alphard

LMI

HYA

Algieba

PYX

CNC

M44

ZENITH

LYN

PUP

Procyon

CMI

NORTH

MON

Pollux

GEM

Castor

Sirius

CMA

M41

M35

AUR

M36 M37

Betelgeuse

M38

ORI

Capella

M79

LEP

Saiph

M42

M1

Bellatrix

Rigel

Aldebaran

TAU

Mira

WEST

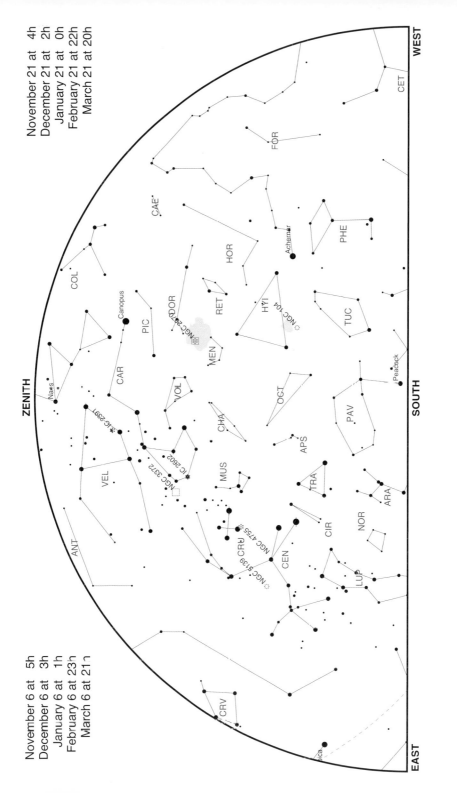

2S

WEST

EAST

SOUTH

ZENITH

November 21 at 4h
December 21 at 2h
January 21 at 0h
February 21 at 22h
March 21 at 20h

November 6 at 5h
December 6 at 3h
January 6 at 1h
February 6 at 23h
March 6 at 21h

CET
FOR
CAE
PHE
Achernar
HOR
COL
RET
TUC
Canopus
PIC
DOR
NGC 2070
MEN
HYI
NGC 104
Naos
CAR
VOL
OCT
Peacock
IC 2391
CHA
PAV
VEL
IC 2602
NGC 3372
MUS
APS
ARA
ANT
CRU
NGC 4755
TRA
NOR
NGC 5139
CEN
CIR
LUP
CRV
Spica

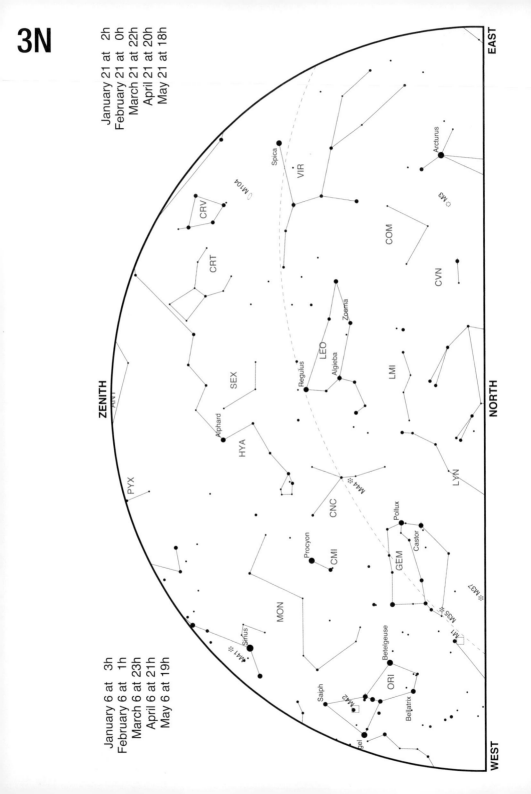

3N

January 21 at 2h
February 21 at 0h
March 21 at 22h
April 21 at 20h
May 21 at 18h

January 6 at 3h
February 6 at 1h
March 6 at 23h
April 6 at 21h
May 6 at 19h

EAST

ZENITH

NORTH

WEST

Spica
VIR
M104
CRV
CRT
SEX
ANT
Alphard
HYA
PYX
M3
Arcturus
COM
CVN
Zosma
LEO
Algieba
Regulus
LMI
LYN
CNC
M44
Procyon
CMI
MON
Pollux
Castor
GEM
M37
M35
M1
Betelgeuse
ORI
Bellatrix
Saiph
M42
Rigel
M41
Sirius

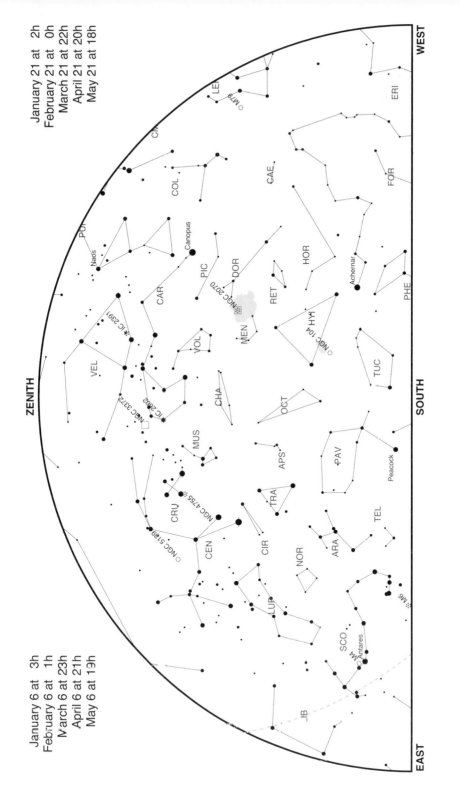

3S

WEST

ZENITH

EAST

SOUTH

LEP
CMA
COL
CAE
ERI
FOR
PHE
PUP
Naos
Canopus
PIC
CAR
DOR
RET
HOR
Achernar
IC 2391
NGC 2070
MEN
HYI
NGC 104
TUC
VEL
VOL
CHA
OCT
PAV
NGC 3372
IC 2602
MUS
APS
Peacock
NGC 4755
CRU
CEN
CIR
TRA
TEL
NGC 5189
NOR
ARA
LUP
SCO
M4
Antares
M6
LIB

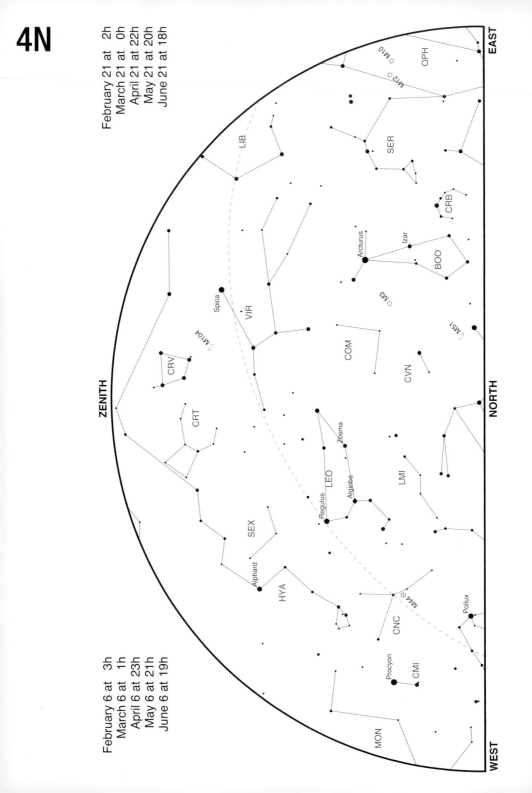

4N

February 21 at 2h
March 21 at 0h
April 21 at 22h
May 21 at 20h
June 21 at 18h

February 6 at 3h
March 6 at 1h
April 6 at 23h
May 6 at 21h
June 6 at 19h

EAST

WEST

NORTH

ZENITH

OPH
M10
M12
SER
CRB
LIB
Arcturus
Izar
BOO
M3
Spica
VIR
M51
COM
M104
CRV
CVN
CRT
Regulus
LEO
Zósma
Algieba
LMI
SEX
Alphard
HYA
M44
CNC
Pollux
Procyon
CMI
MON

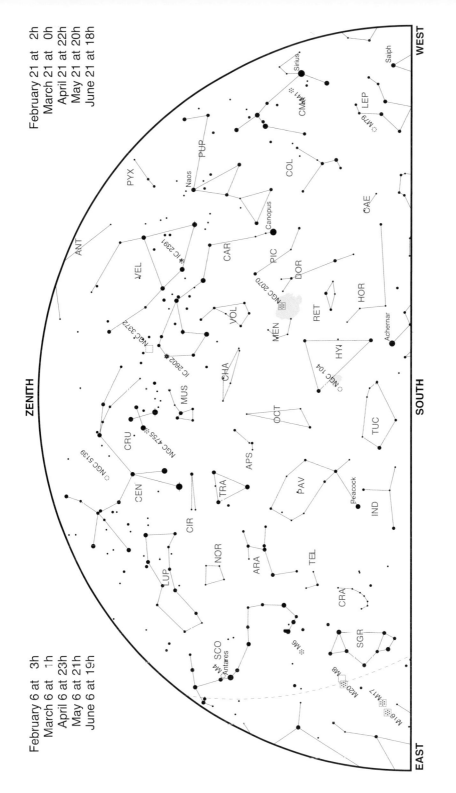

4S

WEST

February 21 at 2h
March 21 at 0h
April 21 at 22h
May 21 at 20h
June 21 at 18h

ZENITH

February 6 at 3h
March 6 at 1h
April 6 at 23h
May 6 at 21h
June 6 at 19h

EAST

SOUTH

Sirius
M41
CMA
Saiph
LEP
M79
COL
PUP
Naos
PYX
CAR
Canopus
CAE
ANT
IC 2391
VEL
PIC
DOR
NGC 2070
MEN
HOR
Achernar
NGC 3372
IC 2602
VOL
RET
HYI
NGC 104
CHA
MUS
OCT
TUC
CRU
NGC 4755
APS
NGC 5139
CEN
TRA
PAV
Peacock
IND
CIR
NOR
ARA
TEL
LUP
SCO
M4
Antares
M6
CRA
SGR
M8
M20
M16
M17

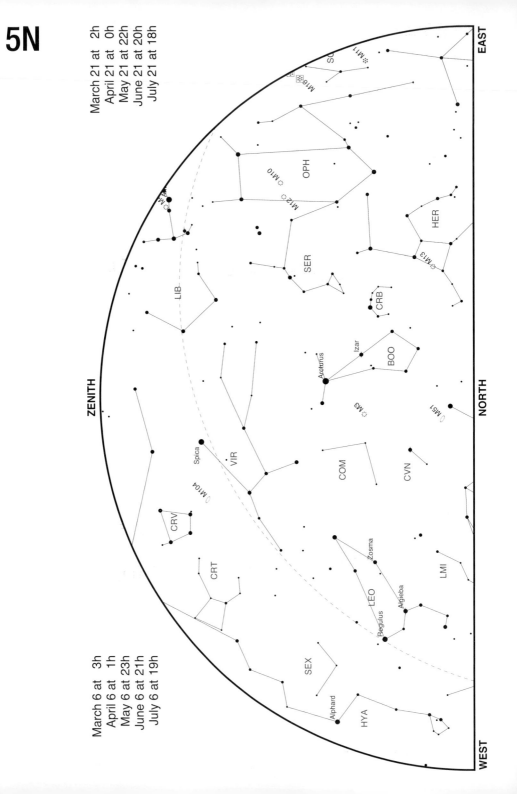

5N

March 21 at 2h
April 21 at 0h
May 21 at 22h
June 21 at 20h
July 21 at 18h

March 6 at 3h
April 6 at 1h
May 6 at 23h
June 6 at 21h
July 6 at 19h

ZENITH

EAST

NORTH

WEST

OPH

SER

HER

CRB

LIB

BOO

Izar

Arcturus

M13

M12

M10

M11

M16

SC

M3

M51

COM

CVN

VIR

Spica

M104

CRV

CRT

SEX

LEO

Zosma

Algieba

Regulus

LMI

Alphard

HYA

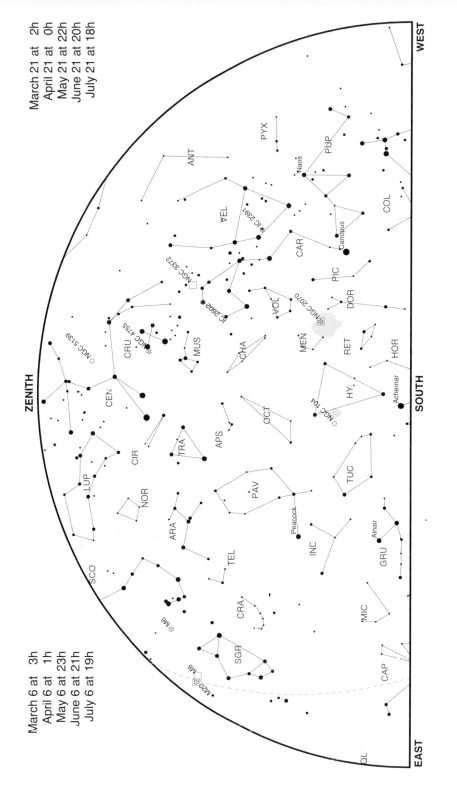

5S

WEST

March 21 at 2h
April 21 at 0h
May 21 at 22h
June 21 at 20h
July 21 at 18h

March 6 at 3h
April 6 at 1h
May 6 at 23h
June 6 at 21h
July 6 at 19h

ZENITH

EAST

SOUTH

PYX
ANT
VEL
IC 2391
CAR
Naos
PUP
Canopus
COL
PIC
NGC 3372
IC 2602
VOL
DOR
NGC 2070
MEN
RET
HOR
CRU
NGC 4755
MUS
CHA
HY.
NGC 104
Achernar
NGC 5139
CEN
CIR
APS
TRA
OCT
LUP
NOR
ARA
PAV
TUC
SCO
TEL
Peacock
INC
GRU
Alnair
M6
CRA
MIC
SGR
M20·M8
CAP

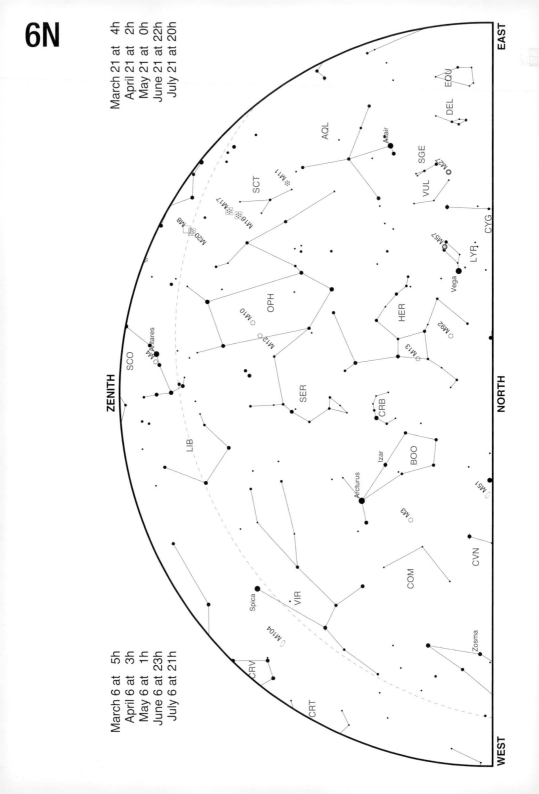

6N

March 6 at 5h
April 6 at 3h
May 6 at 1h
June 6 at 23h
July 6 at 21h

EAST

ZENITH

NORTH

WEST

EQU
DEL
AQL
SGE
VUL
Altair
CYG
LYR
Vega
M57
M27
HER
M92
M13
SCT
M11
M17
M16
M20
M8
OPH
M10
M12
SER
CRB
Antares
SCO
M4
LIB
BOO
Izar
Arcturus
M3
M51
CVN
COM
VIR
Spica
M104
Zosma
CRV
CRT

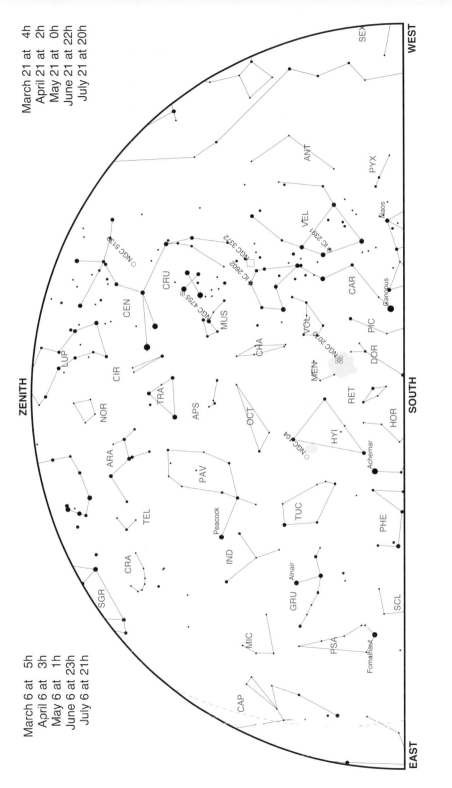

6S

WEST

EAST

SOUTH

ZENITH

March 21 at 4h
April 21 at 2h
May 21 at 0h
June 21 at 22h
July 21 at 20h

March 6 at 5h
April 6 at 3h
May 6 at 1h
June 6 at 23h
July 6 at 21h

SEX

ANT

PYX

VEL

IC 2391

CAR

Naos

Canopus

PIC

VOL

NGC 2070

MEN

DOR

RET

HOR

Achernar

HYI

NGC 104

OCT

CHA

MUS

NGC 4755

IC 2602

NGC 3372

CRU

CEN

NGC 5139

LUP

CIR

NOR

TRA

APS

ARA

TEL

PAV

Peacock

IND

TUC

PHE

SCL

CRA

SGR

MIC

GRU

Alnair

PSA

Fomalhaut

CAP

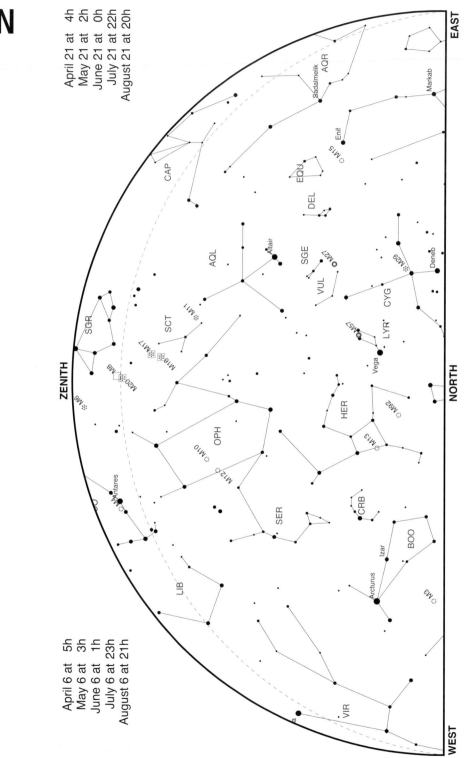

7N

April 21 at 4h
May 21 at 2h
June 21 at 0h
July 21 at 22h
August 21 at 20h

April 6 at 5h
May 6 at 3h
June 6 at 1h
July 6 at 23h
August 6 at 21h

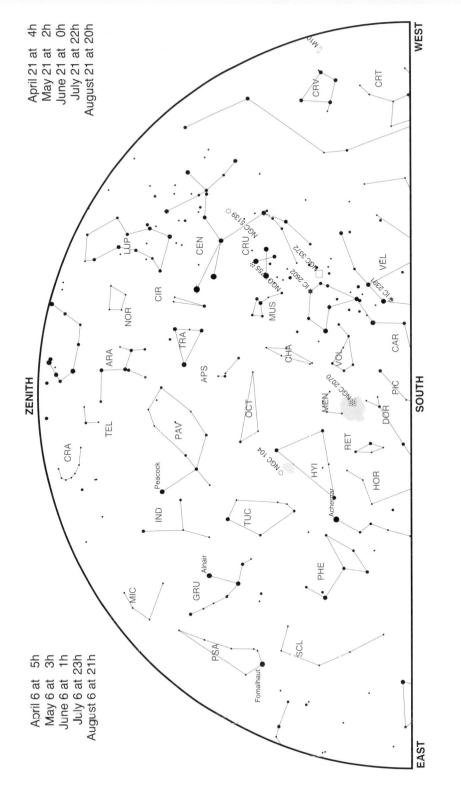

7S

WEST

April 21 at 4h
May 21 at 2h
June 21 at 0h
July 21 at 22h
August 21 at 20h

April 6 at 5h
May 6 at 3h
June 6 at 1h
July 6 at 23h
August 6 at 21h

ZENITH

EAST

SOUTH

CRV
CRT

LUP
CEN
CRU
NGC 5139
NGC 3372
VEL

CIR
NGC 4755
MUS
IC 2602
IC 2391

NOR
CHA
CAR

ARA
TRA
VOL
PIC

APS
MEN
DOR

OCT
NGC 2070

TEL
CRA

PAV
HYI
RET

IND
Peacock
NGC 104
HOR

Achernar

MIC
GRU
Alnair
TUC
PHE

PSA
SCL

Fomalhaut

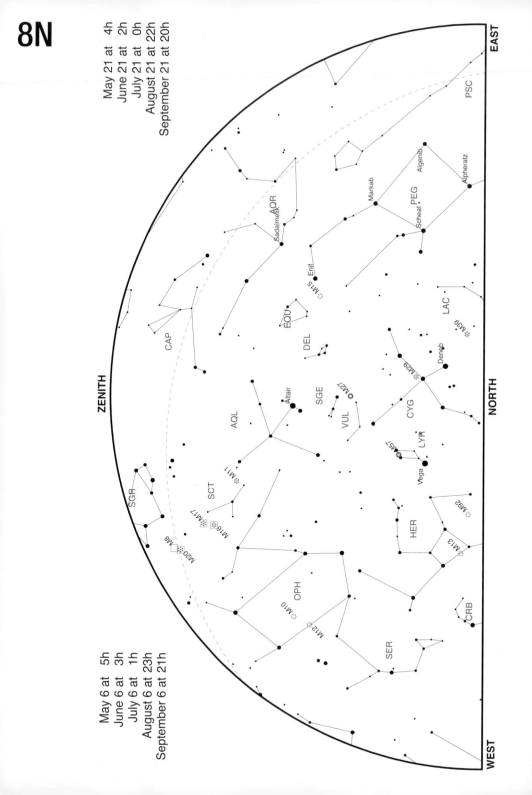

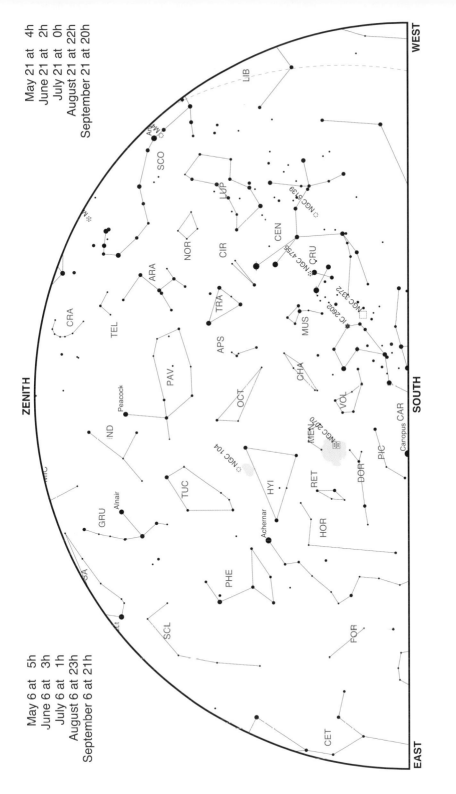

8S

WEST

EAST

SOUTH

ZENITH

May 21 at 4h
June 21 at 2h
July 21 at 0h
August 21 at 22h
September 21 at 20h

May 6 at 5h
June 6 at 3h
July 6 at 1h
August 6 at 23h
September 6 at 21h

LIB
SCO
Antares
M4
LUP
NGC 5139
NOR
CIR
CEN
ARA
TRA
NGC 4755
CRU
TEL
CRA
APS
NGC 4372
IC 2602
MUS
CHA
PAV.
OCT
VOL
IND
Peacock
CAR
Canopus
MEN
NGC 2070
NGC 104
DOR
Alnair
TUC
HYI
RET
PIC
GRU
Achernar
HOR
PHE
SCL
FOR
CET

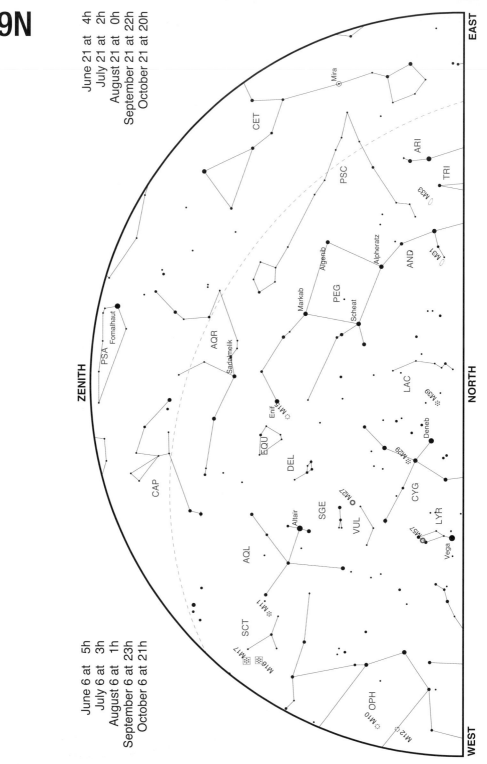

9N

June 6 at 5h
July 6 at 3h
August 6 at 1h
September 6 at 23h
October 6 at 21h

EAST

ZENITH

WEST

NORTH

Mira

CET

PSC

ARI

TRI

M33

AND

M31

Algenib

Alpheratz

Markab

PEG

Scheat

PSA

Fomalhaut

AQR

Sadalmelik

LAC

M39

Enif

M15

EQU

DEL

Deneb

M29

CAP

SGE

M27

VUL

CYG

LYR

Altair

M57

Vega

AQL

M11

SCT

M17

M16

M10

OPH

M12

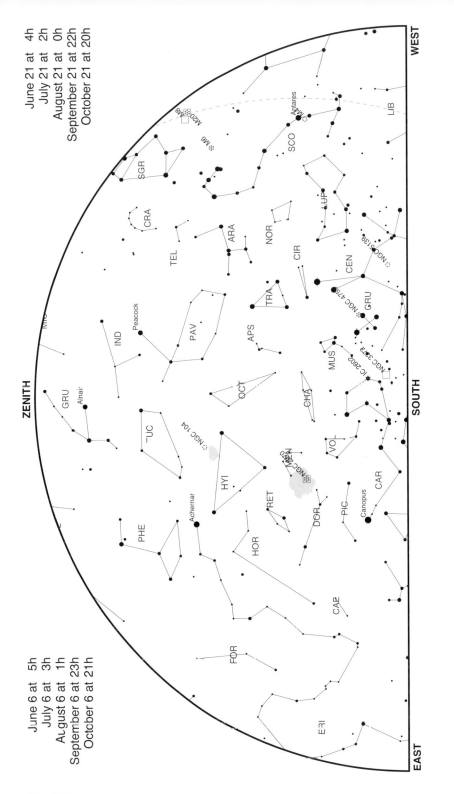

9S

WEST

EAST

SOUTH

ZENITH

June 6 at 5h
July 6 at 3h
August 6 at 1h
September 6 at 23h
October 6 at 21h

SGR
CRA
TEL
ARA
NOR
CIR
TRA
APS
IND
PAV
OCT
Peacock
GRU
Alnair
TUC
HYI
RET
HOR
PHE
Achernar
CAE
FOR
ERI
M20
M8
M6
Antares
SCO
LIB
LUP
CEN
NGC 139
NGC 4755
GRU
MUS
IC 2602
NGC 3372
CHA
VOL
DOR
MEN
NGC 2070
NGC 104
CAR
PIC
Canopus

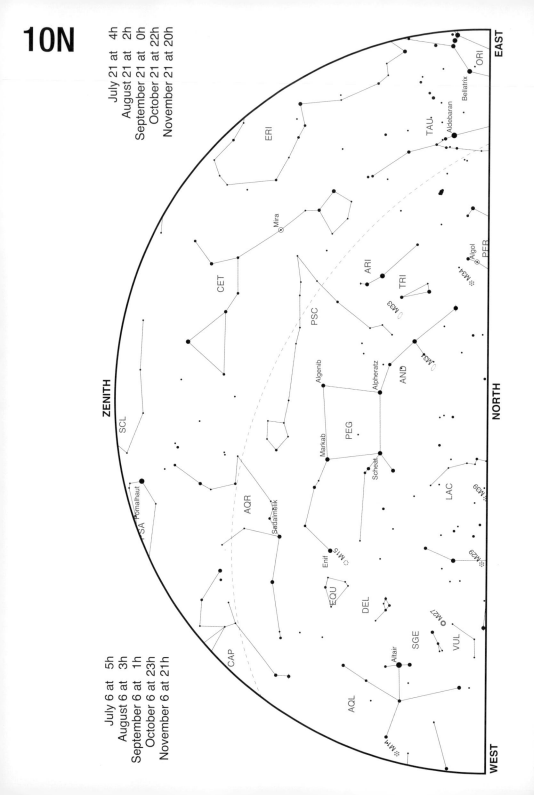

10N

July 6 at 5h
August 6 at 3h
September 6 at 1h
October 6 at 23h
November 6 at 21h

ZENITH

EAST

NORTH

WEST

ORI
Bellatrix
Aldebaran
TAU
Algol
PER
M34
ARI
TRI
M33
M33
ERI
Mira
CET
PSC
AND
Alpheratz
Algenib
PEG
Markab
Scheat
SCL
PsA
Fomalhaut
AQR
Sadalmelik
Skat
M15
Enif
EQU
DEL
SGE
Altair
VUL
M27
M29
M39
LAC
CAP
AQL
M11

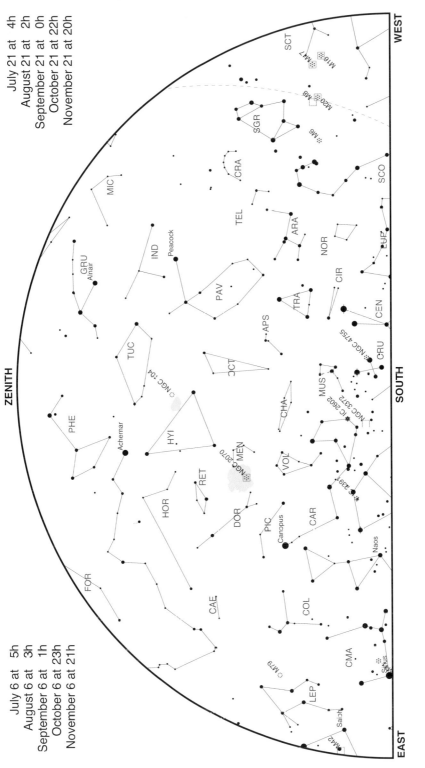

10S

WEST

July 21 at 4h
August 21 at 2h
September 21 at 0h
October 21 at 22h
November 21 at 20h

July 6 at 5h
August 6 at 3h
September 6 at 1h
October 6 at 23h
November 6 at 21h

ZENITH

EAST

SOUTH

SCT

M16
M17

M20
M8

SGR

M6

SCO

CRA

TEL

ARA

LUP

NOR

MIC

IND

Peacock

GRU

Alnair

PAV

APS

TRA

CIR

CEN

NGC 4755

CRU

TUC

NGC 104

OCT

CHA

MUS

IC 2602

NGC 3372

PHE

Achernar

HYI

MEN

NGC 2070

VOL

3372

RET

HOR

DOR

PIC

Canopus

CAR

FOR

CAE

COL

Naos

CMA

Sirius

M79

LEP

Saiph

Rigel

M42

11N

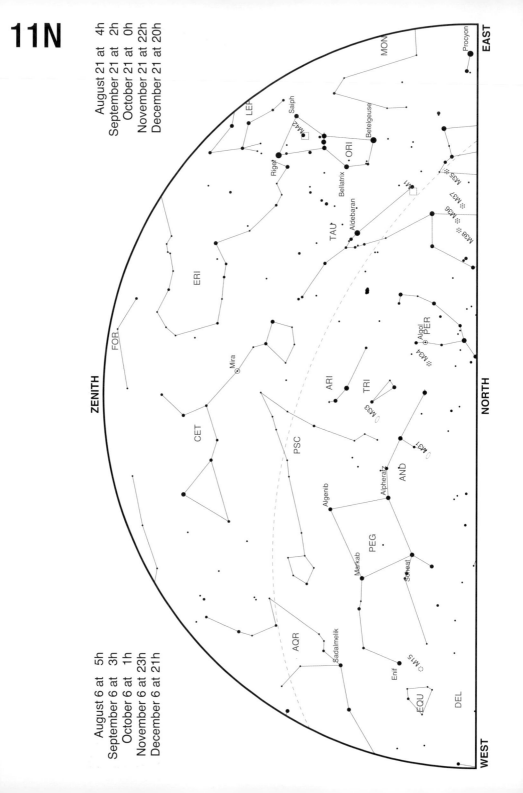

August 21 at 4h
September 21 at 2h
October 21 at 0h
November 21 at 22h
December 21 at 20h

August 6 at 5h
September 6 at 3h
October 6 at 1h
November 6 at 23h
December 6 at 21h

EAST

WEST

NORTH

ZENITH

MON
Procyon
LEP
Saiph
M42
ORI
Betelgeuse
Rigel
Bellatrix
Aldebaran
TAU
ERI
FOR
Mira
CET
M1
M35
M37
M36
M38
Algol
PER
M34
ARI
TRI
M33
M31
AND
Alpheratz
Algenib
PSC
PEG
Markab
Scheat
AQR
Sadalmelik
Enif
M15
EQU
DEL

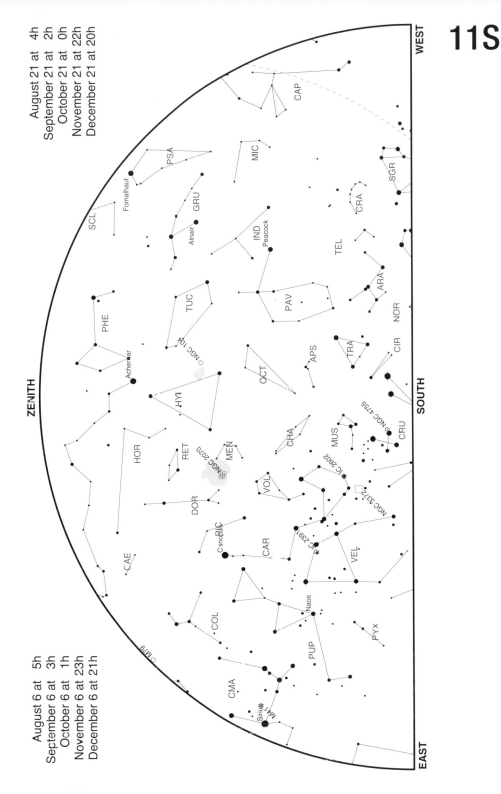

11S

WEST

ZENITH

SOUTH

EAST

CAP

PSA
Fomalhaut

SCL

MIC

SGR

CRA

GRU
Alnair

IND
Peacock

TEL

ARA

NOR

PHE
Achernar

TUC

NGC 104

PAV

CIR

HYI

OCT

APS

TRA

HOR

RET

MEN
NGC 2070

CHA

MUS

NGC 4755

CRU
IC 2602
NGC 3372

DOR

VOL

CAE

PIC
Canopus

CAR

IC 2391

VEL

PYX

COL

PUP
Naos

CMA
Sirius
M41

M/W

12N

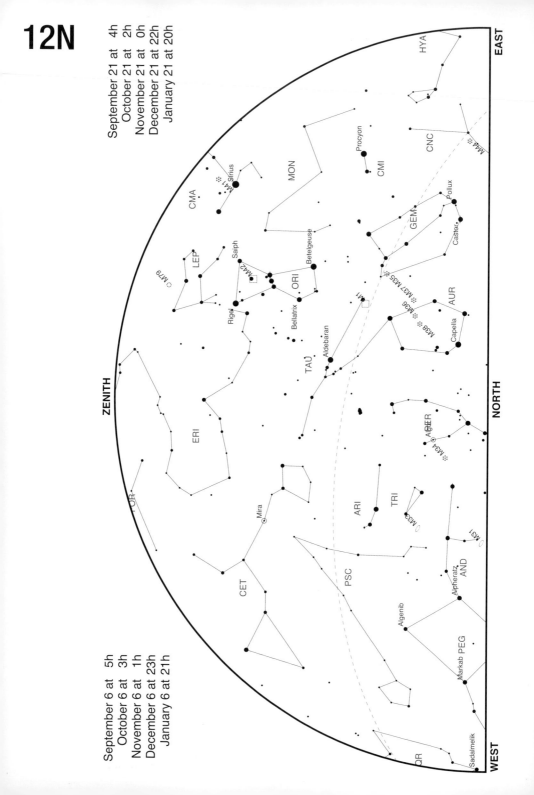

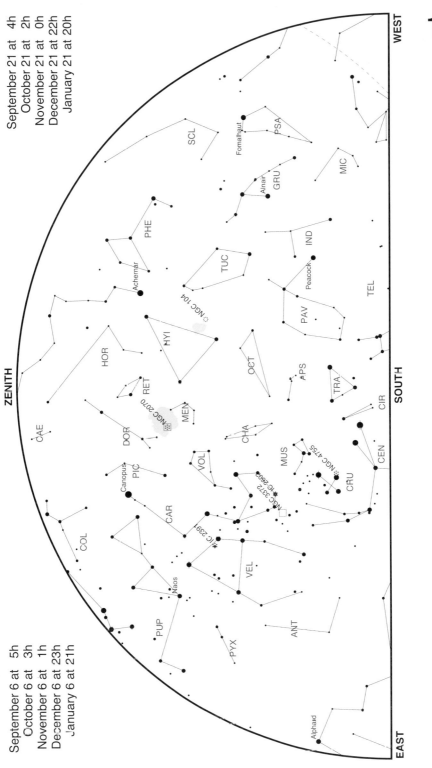

12S

WEST

ZENITH

SOUTH

EAST

SCL

Fomalhaut

PSA

Alnair

GRU

MIC

PHE

Achernar

TUC

IND

Peacock

TEL

NGC 104

HYI

PAV

HOR

RET

OCT

APS

TRA

CAE

DOR

MEN

NGC 2070

CHA

CIR

CEN

VOL

MUS

NGC 4755

PIC

NGC 3372

IC 2602

CRU

Canopus

CAR

IC 2391

COL

VEL

Naos

ANT

PUP

PYX

Alphad

The Planets in 2022

Lynne Marie Stockman

Mercury never strays far from the Sun and thus is a challenging object to find in twilit skies. Its best evening apparition for northern temperate observers is April–May whilst southern hemisphere viewers must wait until mid-July through late September. Mercury's best morning shows are in February and March for those south of the equator and late September to early November for those farther north. Mercury is brightest near superior conjunction and faintest at inferior conjunction so evening apparitions are characterised by a bright start followed by a steady dimming whilst morning apparitions start faintly and end brightly. Mercury passes near the Pleiades (M45) in April and is found just north of Regulus (α Leonis) in August. Its closest encounter with another planet occurs in March when it meets Saturn. Mercury is occulted twice by the Moon, in October and November. *Apparition diagrams showing the position of Mercury above the eastern and western horizons can be found throughout the Sky Notes.*

Venus spends the first few days of 2022 as the evening star but it soon passes into the morning sky where it remains for ten months, returning to evening twilight in November. This is an excellent morning apparition for southern hemisphere astronomers and it is also quite good for those living near Earth's equator. However, the bright planet never gains much altitude when viewed from northern temperate zones, remaining within 20° or so of the horizon. Greatest elongation west occurs in March, with inferior conjunction taking place in January and superior conjunction in October. It passes very close to both Neptune and Jupiter in late April. *Apparition diagrams showing the position of Venus above the eastern and western horizons can be found in the January and October Sky Notes respectively.*

The **Moon** makes a series of close passes by several first-magnitude stars this year. Pollux (β Geminorum) is approached each month, with the smallest separation (1.7°) occurring in November. The distance between the Moon and Aldebaran (α Tauri) dips to below 3° only in June but our satellite closes in on Antares (α Scorpii) all year, with the distance between the two bodies finally becoming less than 3° in August; monthly occultations of this bright star will begin next year. The open star cluster M45 (Pleiades) is also a lunar target in the latter part of the year. A number of occultations of planetary bodies takes place, with Mercury, Venus, Mars and Uranus all disappearing behind the Moon's disk at various times of the year.

Mars begins the year in Ophiuchus but soon moves into Sagittarius before progressing across the zodiac (with a brief foray into Cetus in June) to Taurus where it arrives in August. Mars encounters every other planet except Mercury, approaching Venus twice (February and March), Saturn (April), Neptune and Jupiter (May) and finally Uranus (August). It is occulted by the Moon three times but only the last occultation, in December during the planet's opposition, is readily visible. The red planet spends most of the year in the morning sky, not rising before midnight until July for northern temperate latitudes and September–October for those in the southern hemisphere. *A finder chart showing the position of Mars in the first half of the year can be found in the January Sky Notes. A second finder chart showing the position of Mars in the last half of the year can be found in the December Sky Notes. Background stars are shown to magnitude +5.5.*

Jupiter starts out as the dominant planetary feature of the evening sky. It disappears in the west in late February, undergoing conjunction in early March and reappearing in the morning sky later that month. A rare planetary conjunction with Neptune takes place in the dawn skies of April and the largest planet in the solar system teams up with the brightest one at the end of that month for a spectacular pairing in the morning skies. The Moon makes a series of reasonably close passes later in the year but never gets close enough to occult the gas giant. Opposition occurs in late September. Jupiter begins the year in Aquarius but moves to Pisces in April before spending two months in Cetus. It eventually returns to Pisces in September where it remains for the rest of the year. *A finder chart showing the position of Jupiter throughout 2022 can be found in the September Sky Notes. Background stars are shown to magnitude +5.5.*

Saturn continues its slow journey across the watery constellation of Capricornus. Briefly visible low in the west at the beginning of the year, Saturn undergoes conjunction in early February and reappears in the dawn sky later that month. The three rocky planets all come to call, Mercury and Venus in March and Mars in April, but the Moon never draws near and the ringed planet does not pass any bright stars. The rings close to a minimum tilt of just over 12° in early June before reopening temporarily and the planet comes to opposition in August when it reaches almost zero magnitude. It is an evening sky object for the remainder of the year. *A finder chart showing the position of Saturn throughout 2022 can be found in the August Sky Notes. Background stars are shown to magnitude +5.5.*

Uranus remains in Aries all year, passing near the sixth-magnitude star Omicron (o) Arietis in April and being occulted by an eclipsed Full Moon in November. Uranus is visible in the evening sky at the outset of 2022 and undergoes solar conjunction in May. It reappears in the morning sky, rising earlier every day and moving in the evening sky again by August or September, depending upon the observer's latitude. Opposition occurs in November. Except for January, the Moon occults Uranus every month this year. *A finder chart showing the position of Uranus throughout 2022 can be found in the November Sky Notes. Background stars are shown to magnitude +8.0.*

Neptune spends the year zigzagging between Aquarius and Pisces. Visible in the evening sky in January, it reaches conjunction with the Sun in mid-March. April brings a relatively rare conjunction with Jupiter with the two planets appearing just 0.1° apart in the morning sky; this event last occurred in 2009. Neptune is passed even more closely by Venus a few days later. Opposition takes place in mid-September and Neptune returns to the evening sky for the remainder of the year. *A finder chart showing the position of Neptune throughout 2022 can be found in the September Sky Notes. Background stars are shown to magnitude +10.0.*

Some Events in 2022

January	1	Saturn	Maximum ring opening (17.6°)
	3/4	Earth	Quadrantid meteor shower (ZHR 120)
	4	Earth	Perihelion (0.9833 au)
	7	Mercury	Greatest elongation east (19.2°)
	9	Venus	Inferior conjunction (evening → morning)
	13	7 Iris	Opposition (magnitude +7.7 in Gemini)
	17	Full Moon	Smallest apparent diameter of the year (0.4967°)
	23	Mercury	Inferior conjunction (evening → morning)
	26	Mars	0.5° north of M8 (Lagoon Nebula)
	30	Uranus	East quadrature

February	4	Saturn	Conjunction
	5	Mars	0.2° north of M22
	7	Moon, Uranus	Lunar occultation (82° from the Sun)
	16	Mercury	Greatest elongation west (26.3°)

March	5	Jupiter	Conjunction
	7	Moon, Uranus	Lunar occultation (55° from the Sun)
	10	Moon	Nearest apogee of the year (404,300 km)
	13	Neptune	Conjunction
	20	Venus	Greatest elongation west (46.6°)
	20	Earth	Equinox
	21	Venus	Theoretical dichotomy
	23	Moon	Farthest perigee of the year (369,800 km)
	28	Venus, Saturn	Planetary conjunction (1.2° apart, 46° from the Sun)

April	2	Mercury	Superior conjunction (morning → evening)
	3	Moon, Uranus	Lunar occultation (29° from the Sun)
	5	Mars, Saturn	Planetary conjunction (0.3° apart, 53° from the Sun)
	12	8 Flora	Opposition (magnitude +9.8 in Virgo)
	12	Jupiter, Neptune	Planetary conjunction (0.1° apart, 29° from the Sun)
	22/23	Earth	Lyrid meteor shower (ZHR 18)
	27	Venus, Neptune	Planetary conjunction (0.01° apart, 43° from the Sun)
	29	Mercury	Greatest elongation east (20.6°)
	30	Sun, New Moon	Partial solar eclipse
	30	Venus, Jupiter	Planetary conjunction (0.2° apart, 43° from the Sun)

May	1	Moon, Uranus	Lunar occultation (4° from the Sun)
	5	Uranus	Conjunction
	6/7	Earth	Eta Aquariid meteor shower (ZHR 30)
	15	Saturn	West quadrature
	16	Earth, Full Moon	Total lunar eclipse
	21	Mercury	Inferior conjunction (evening → morning)
	27	Moon, Venus	Lunar occultation (38° from the Sun)
	28	Moon, Uranus	Lunar occultation (21° from the Sun)
	29	Mars, Jupiter	Planetary conjunction (0.6° apart, 65° from the Sun)

June	1	Saturn	Minimum ring opening (12.3°)
	9	Earth	Tau Herculid meteor shower (ZHR varies)
	16	Neptune	West quadrature
	16	Mercury	Greatest elongation west (23.2°)
	21	Earth	Solstice
	22	Moon, Mars	Lunar occultation (71° from the Sun)
	24	Moon, Uranus	Lunar occultation (46° from the Sun)
	29	Jupiter	West quadrature
	29	Moon	Farthest apogee of the year (406,600 km)

July	4	Earth	Aphelion (1.107 au)
	13	Moon	Nearest perigee of the year (357,300 km)
	13	Full Moon	Largest apparent diameter of the year (0.5572°)
	16	Mercury	Superior conjunction (morning → evening)
	21	Moon, Mars	Lunar occultation (77° from the Sun)
	22	Moon, Uranus	Lunar occultation (71° from the Sun)
	28/29	Earth	Delta Aquariid meteor shower (ZHR 20)

August	11	Uranus	West quadrature
	12/13	Earth	Perseid meteor shower (ZHR 80)
	14	Saturn	Opposition (magnitude +0.3 in Capricornus)
	18	Moon, Uranus	Lunar occultation (97° from the Sun)
	22	4 Vesta	Opposition (magnitude +5.7 in Aquarius)
	27	Mars	West quadrature
	27	Mercury	Greatest elongation east (27.3°)

September	7	3 Juno	Opposition (magnitude +7.8 in Aquarius)
	8	5 Astraea	Opposition (magnitude +10.9 in Aquarius)
	10	Full Moon	'Harvest Moon'
	14	Moon, Uranus	Lunar occultation (123° from the Sun)
	16	Neptune	Opposition (magnitude +7.8 in Aquarius)
	23	Earth	Equinox
	23	Mercury	Inferior conjunction (evening → morning)
	26	Jupiter	Opposition (magnitude −2.9 in Pisces)

	8	Mercury	Greatest elongation west (18.0°)
	8/9	Earth	Draconid meteor shower (ZHR 10)
	10	Earth	Southern Taurid meteor shower (ZHR 5)
	12	Moon, Uranus	Lunar occultation (151° from the Sun)
October	21/22	Earth	Orionid meteor shower (ZHR 20)
	22	Venus	Superior conjunction (morning → evening)
	24	Moon, Mercury	Lunar occultation (10° from the Sun)
	25	Sun, New Moon	Partial solar eclipse
	25	Moon, Venus	Lunar occultation (1° from the Sun)

	8	Earth, Full Moon	Total lunar eclipse
	8	Moon, Uranus	Lunar occultation (179° from the Sun)
	8	Mercury	Superior conjunction (morning → evening)
	9	Uranus	Opposition (magnitude +5.6 in Aries)
November	11	Saturn	East quadrature
	12	Earth	Northern Taurid meteor shower (ZHR 5)
	12	27 Euterpe	Opposition (magnitude +8.7 in Aries)
	17/18	Earth	Leonid meteor shower (ZHR varies)
	24	Moon, Mercury	Lunar occultation (9° from the Sun)

	1	Mars	Minimum geocentric distance (0.5447 au)
	5	Moon, Uranus	Lunar occultation (153° from the Sun)
	8	Moon, Mars	Lunar occultation (180° from the Sun)
	8	Mars	Opposition (magnitude −1.9 in Taurus)
December	13/14	Earth	Geminid meteor shower (ZHR 75+)
	14	Neptune	East quadrature
	21	Mercury	Greatest elongation east (20.1°)
	21	Earth	Solstice
	22/23	Earth	Ursid meteor shower (ZHR 10)
	22	Jupiter	East quadrature

Note that the dates quoted in the calendar are based on UT and that events listed may occur one day before or after the given date depending on the observer's time zone.

The entries relating to meteor showers state the *estimated* date of peak shower activity (maximum). The figure quoted in parentheses in column 4 alongside each meteor shower entry is the expected Zenith Hourly Rate (ZHR) for that particular shower at maximum. For a more detailed explanation of ZHR, and for further details of the individual meteor showers listed here, please refer to the article *Meteor Showers in 2022* located elsewhere in this volume. For more on each of the eclipses occurring during the year, please refer to the information given in *Eclipses in 2022*.

Phases of the Moon in 2022

New Moon	First Quarter	Full Moon	Last Quarter
2 January	9 January	17 January	25 January
1 February	8 February	16 February	23 February
2 March	10 March	18 March	25 March
1 April	9 April	16 April	23 April
30 April	9 May	16 May	22 May
30 May	7 June	14 June	21 June
29 June	7 July	13 July	20 July
28 July	5 August	12 August	19 August
27 August	3 September	10 September	17 September
25 September	3 October	9 October	17 October
25 October	1 November	8 November	16 November
23 November	30 November	8 December	16 December
23 December	30 December		

Eclipses in 2022

There are a minimum of four eclipses in any one calendar year, comprising two solar eclipses and two lunar eclipses. Most years have only four, as is the case with 2022, although it is possible to have five, six or even seven eclipses during the course of a year. It is important to note that the times quoted for each event below refer to the start, maximum and ending of the eclipse on a global scale rather than with reference to specific locations. As far as lunar eclipses are concerned, these events are visible from all locations that happen to be on the night side of the Earth. Depending on the exact location of the observer, the entire eclipse sequence may be visible, although for some the Moon will be either rising or setting while the eclipse is going on.

The first eclipse of the year is the partial solar eclipse of 30 April which will be visible throughout most of the southeast Pacific Ocean, southern South America and parts of Antarctica, with the best views available from Argentina where the coverage will be a little over 50%. The eclipse begins at 18:45 UT and ends at 22:38 UT with maximum eclipse occurring at 20:41 UT.

On 16 May there will be a total lunar eclipse, the whole or parts of which will be visible from North America (except Alaska, the Yukon and the islands of Nunavut), Central and South America, southern Greenland and Iceland, the Atlantic Ocean and Antarctica, parts of western and southern Europe, and western Africa.From the British Isles, the Moon sets whilst it is still fully immersed in the Earth's umbra (full shadow). The eclipse commences when the Moon enters the Earth's penumbra (partial shadow) at 01:32 UT and ends at 06:51 UT. The Moon begins to enter the Earth's umbra at 02:28 UT, with full eclipse lasting from 03:29 UT to 04:54 UT and maximum eclipse occurring at 04:11 UT. The Moon leaves the umbra at 05:55 UT.

There will be a partial solar eclipse on 25 October which will be visible from Europe, south and west Asia, northern and eastern Africa and the northern Atlantic Ocean, with the best views being from parts of western Russia and Kazakhstan. From the

British Isles, the eclipse will have a magnitude between 0.18 and 0.40, with Scotland and northern England seeing the deepest eclipse. The eclipse begins at 08:58 UT and ends at 13:02 UT with maximum eclipse at 11:00 UT.

A total lunar eclipse will take place on 8 November, the entirety or parts of which will be visible throughout eastern Russia, the Bering Strait, North America and the Arctic, northern Greenland, Japan, most of South America, the Indian Ocean, New Zealand and Australia, and most of the Pacific Ocean. The eclipse commences when the Moon enters the Earth's penumbra (partial shadow) at 08:02 UT and ends at 13:56 UT. The Moon begins to enter the Earth's umbra (full shadow) at 09:09 UT, with full eclipse lasting from 10:17 UT to 11:41 UT and maximum eclipse occurring at 10:59 UT. The Moon leaves the umbra at 12:49 UT. During this event, Uranus is occulted by the eclipsed Full Moon. For further details see the entry for Uranus in the Monthly Sky Notes for November.

Monthly Sky Notes and Articles

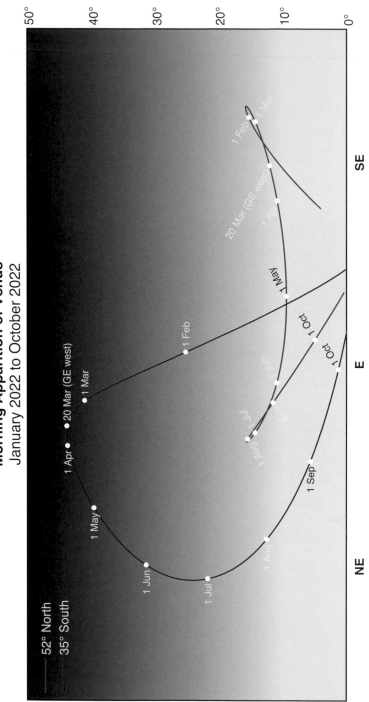

Morning Apparition of Venus
January 2022 to October 2022

52° North
35° South

January

New Moon: 2 January
Full Moon: 17 January

MERCURY is climbing higher above the western horizon as the New Year commences, reaching greatest elongation east (19.2°) on 7 January. It soon heads back toward the Sun, with inferior conjunction occurring on 23 January, only to reappear in the east at dawn at the end of the month. The tiny planet crosses the ecliptic at its ascending node on 11 January, embarks on retrograde motion on 14 January and reaches the first of four perihelia a day later. Mercury exhibits a wide range of brightnesses this month, starting at magnitude −0.7, dimming to sixth magnitude around inferior conjunction and then recovering to +1.3 in the morning sky as January draws to a close.

VENUS opens the year low in the west, and then disappears after a few days to undergo inferior conjunction on the ninth. It swiftly reappears as the morning star where it will reign until mid-October. Look for it in the east before sunrise. Perihelion occurs on 23 January. Having entered 2022 in retrograde, Venus returns to direct motion on 29 January.

EARTH reaches the point in its orbit where it is nearest to the Sun on 4 January. This year's perihelion distance is 0.9833 au or 147,100,000 km. The Quadrantid meteor shower peaks at around the same time; details are in *Meteor Showers in 2022*. The Full Moon of 17 January is the 'smallest' of the year, with our satellite exhibiting an apparent lunar diameter of only 1787 arc-seconds (compared to an average of about 1860 arc-seconds). The 'largest' Full Moon will occur in July.

MARS leaves the non-zodiacal constellation of Ophiuchus for Sagittarius on 19 January. Currently a morning sky object after last October's conjunction, Mars rises around two hours before the Sun and is most easily observed from the southern hemisphere. The magnitude +1.5 planet passes just north of M8, the Lagoon Nebula, on 26 January and attains it maximum southerly declination the day after. The waning crescent Moon approaches to within 2.4° on 29 January.

Mars
January to June 2022

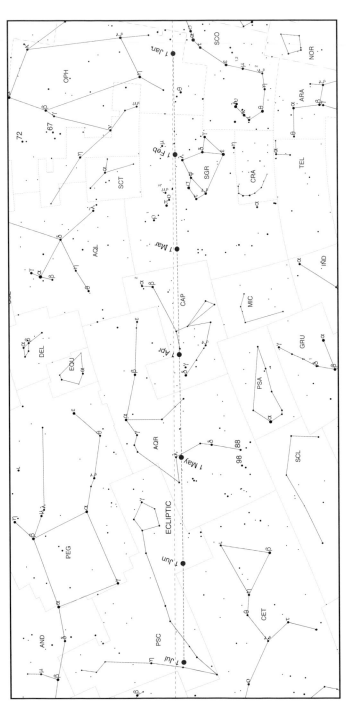

Background stars are shown to magnitude +5.5.

JUPITER is an evening sky object, setting mid- to late-evening, depending on the observer's latitude. Shining brilliantly at magnitude −2.1, it is the brightest star-like object in the constellation of Aquarius. It is at its southernmost declination of the year on 1 January.

SATURN is only a month away from solar conjunction and thus is found low in the west, setting during evening twilight. Located in the constellation of Capricornus, the first-magnitude planet is at its maximum declination south for the year on the first day of January. Its rings are closing as it approaches a ring-plane-crossing event in 2025; the maximum ring opening of 2022 is 17.6° which also occurs at the beginning of the month.

URANUS spends the entirety of 2022 in the faint constellation of Aries. Like Venus, it is in retrograde as the year opens. Uranus is visible in the evening sky, setting after midnight. The waxing gibbous Moon is only 1.3° south of the sixth-magnitude planet on 11 January; this is the only month this year that a lunar occultation of the green ice giant does not take place. Uranus reaches its maximum declination south on 17 January and returns to direct motion the following day. East quadrature, when a superior planet appears 90° away from the Sun, occurs on the penultimate day of the month.

NEPTUNE begins 2022 in the constellation of Aquarius, at its maximum declination south for the year. At magnitude +7.9, it is a telescopic object only and is best viewed early this month from the dark moonless winter skies of the northern hemisphere.

Joseph Thomas Ward
Shepherd Astronomer of New Zealand

John McCue

The hands of Joseph Thomas Ward mastered many artisan skills, but his mind swept like the sirocco across the universe, bringing the stars and planets to the people of New Zealand, and the town of Whanganui in particular. When Joseph died, aged sixty five, Sir Robert Stout, the thirteenth prime minister of New Zealand wrote of him: "I know of no man in the Dominion of greater intellect and reasoning power than the late Mr. Ward."

Joseph was born in Chelsea, England, on 25 January 1862, and though he was educated for the Catholic priesthood, he headed out to sea as a merchant sailor, settling in New Zealand before his teenage years were over. There,

Joseph Thomas Ward with the Cooke refractor. (Whanganui Astronomical Society)

in Marlborough, he worked as a shepherd and shearer, then continued honing his practical skills as a saddler in Wellington. His talented hands later produced pioneering reflecting telescopes in New Zealand, the largest of which had a mirror over twenty inches in diameter. He married Ada Evelyn Wright in 1894 in Wellington, moving to Whanganui two years later where he opened a lending library and bookshop. Like many astronomers before him, he was a talented musician (and poet), teaching the violin, and naming one of his sons William Herschel, after the renowned astronomer who was a professional organist in Bath in Somerset, England, many years before him.

The Cooke 9½ inch refractor, with Whanganui Astronomical Society member Ray Wright. (John McCue)

His early interest in astronomy began to resurface in the dark skies of Whanganui on the North Island. I travelled to this town in 2016 to visit Frank Gibson, a school friend, and saw for myself the magnificent vistas of the night sky – they must have been an irresistible draw for a man of Ward's enquiring nature. Indeed, Ward bought a 4½-inch refractor when he was 36, and when a bright comet appeared two years later, the people of Whanganui arrived in crowds to observe it through his telescope. Following the success of Ward-organised public lectures, the Whanganui Astronomical Society was formed, with Joseph elected as its president. With great determination and passion, he searched for a major telescope as the society's focal point, locating a 9½-inch Cooke refractor in England for £450. It belonged to a Mr. Isaac Fletcher of Clifton and Crossbarrow Collieries in Westmoreland[1] who used it to sketch Saturn. This Cooke instrument was generously shipped out free of charge by the G. D. Tyser & Co. shipping line to its

---

1. *The Victorian Amateur Astronomer*, by Allan Chapman, John Wiley and Sons (1998).

new home in the southern hemisphere. Thomas Cooke of York, the nineteenth-century telescope maker, became Europe's most renowned constructor of these achromatic refractors. Such an object glass comprised two lenses, of course, the second needed to cancel out the chromatic aberration of the first. The young Thomas Cooke came to appreciate the real secret, which was to measure the refractive index of each glass, then calculate the exact geometrical nature of the curves to be ground into each of the four optical surfaces. The process demanded scientific mastery throughout. The Ward Refractor, as it became known, was undoubtedly the finest telescope of its kind in New Zealand when it finally docked at Whanganui.

Joseph Ward led a society delegation to the local authority for a site where this masterpiece of a telescope could be located, and he was granted a place in Cooks Gardens. He visualised the observatory and it was duly designed and built by Mr. A. Atkins at a cost of £290 and opened in 1903 by the Prime Minister, Richard Seddon. Ward quickly realised the potential for serious astronomical research with this telescope. He observed sunspots, comets and Mars, though his main pursuit was the observation of double stars, of which he discovered 108. They are internationally recognised under the designation NZO in the Washington Double

The Ward observatory in Cooks Gardens, Whanganui. (Frank Gibson)

Star Catalogue. This fact enthralled me and I determined to observe, during my 2016 visit, some of Ward's doubles through the very telescope he used to find them.

On 1 July 2016, the society held an entertaining public session when many youngsters were inspired by spectacular views of Mars, Jupiter and Saturn. The society members were carrying forward Joseph's real aim in his life, which was to open the eyes of the public to the wide universe beyond their everyday lives. When Ross Skilton (the current society president) and I were then alone, we computed the current celestial coordinates of the well-placed double NZO50, from its J2000 values, turned the original brass setting circles to those current values, inserted an appropriate brass eyepiece, also original, and behold, there was the double star in the field of view! It was a moment to remember. The hand-drawn sketch made at the telescope eyepiece is shown here. Back in the UK, at a subsequent meeting of the British Astronomical Association, I met Terry Evans, who has a remote telescope near Moorook, South Australia. He kindly agreed to image the Ward doubles which formed part of my observational plans with the Ward refractor, and his image of NZO50 is shown here. Using Aladin, the free professional-

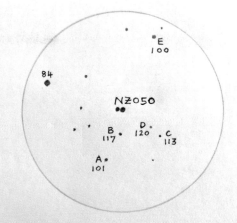

1/7/16   NZO 50   Whanganui, NZ
9½" Cooke refractor
at Joseph Ward's public observatory
Double star magnitudes 9.90 and 10.57
          at distances 82 Ly and 79 Ly
Separation 9.5", position angle 4° (WDSC 2015)
Field of view of eyepiece 20'.

Sketch of NZO50 made at the eyepiece of the Cooke 9½-inch refractor. The stars marked ABCDE are shown in order to match the sketch with the image shown below. The magnitudes are shown without the decimal point. (John McCue)

standard software for analysing images, I measured the separation of NZO50 as 9.4″ (error 0.1″) and its position angle as 5.4° (error 0.3°). The Washington Double Star Catalogue values are given in this sketch. The Washington Double Star Catalogue is mirrored in the website **stelledoppie.it** which is useful for its ability to search the huge catalogue of double stars which has now accumulated. Using the photometry tool in Aladin, the magnitudes were determined as 9.85 (error 0.10) for the primary and 10.50 (error 0.02) for the secondary. NZO50 is not a gravitationally bound pair of stars in orbit around each other, but merely a double star in the same line of sight.

Joseph died on 4 January 1927, and was survived by his wife, four sons and three daughters. His son William Herschel Ward was honorary director of the Whanganui Observatory from 1927 until 1959. In his later years Joseph turned away from religion, as implied in the plaque to his memory on the observatory wall – "Mankind was his religion, the world his country". Joseph Ward was described as gentle, kindly and generous to a fault.

Image of NZO50 taken by Terry Evans with the 8-inch apochromatic refractor housed in his remote observatory near Moorook, South Australia. (Terry Evans)

February

New Moon: 1 February
Full Moon: 16 February

MERCURY exhibits its best morning apparition for southern hemisphere observers this month and next, soaring to well over 20° above the eastern horizon. However, those living in northern temperate latitudes will struggle to see the tiny planet as it barely rises to even 10° in altitude. Mercury brightens steadily all month, beginning at magnitude +1.3 and ending up at −0.1. It finishes retrograde motion on the third day of February. The largest of Mercury's greatest elongations west occurs on 16 February when the planet is 26.3° away from the Sun. Two days later, on 18 February, it reaches the descending node in its orbit, crossing the ecliptic plane from north to south, and on the last day of the month, goes through aphelion for the first time this year.

Morning Apparition of Mercury
23 January to 2 April

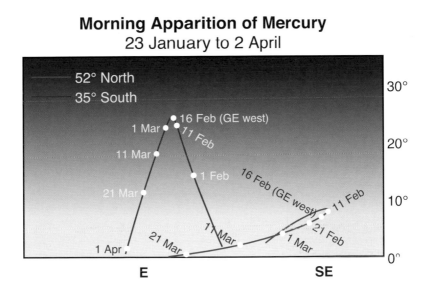

VENUS is at its brightest during this morning apparition when it reaches magnitude −4.6 in early February. It appears 6.2° north of Mars on 16 February when the two objects are 41° west of the Sun but the two planets will have a closer encounter in March. The morning star remains fairly low for observers in northern temperate latitudes but equatorial and southern viewers will see the bright planet vault high above the eastern horizon this month.

EARTH witnesses an occultation of Uranus by the Moon on the seventh day of the month but you will need to charter a boat to the south Atlantic to see it. However, this is just the first in a series of occultations that continue into next year so you may get your chance later.

MARS is moving through Sagittarius this month and passes just 0.2° north of the globular cluster M22 on 5 February. It has a rather distant planetary conjunction with the morning star on 16 February when the two bodies are just over 6° apart in the sky. Even though it is a respectable magnitude +1.4, Mars is much fainter than Venus. Mars is most easily observed from southern latitudes where it rises well in advance of the dawn.

JUPITER is visible low in the west at the beginning of the month but with conjunction looming in early March, it soon vanishes in the glare of the setting Sun.

SATURN is at conjunction on 4 February and is lost to view for much of the month. For southern hemisphere early risers, the planet reappears low in the east before sunrise from mid-month but observers in northern temperate latitudes will have to wait until March to reacquire Saturn in morning skies.

URANUS is occulted by the waxing crescent Moon on 7 February but this event is only visible from the southern Atlantic Ocean. An evening sky object in Aries, sixth-magnitude Uranus is best seen from the long dark winter skies of the northern hemisphere.

NEPTUNE continues its slow journey across Aquarius. It is visible in the evening sky after sunset but is disappearing in the western twilight as it draws nearer to its date with the Sun next month. It is best viewed from the northern hemisphere but point your telescope west as soon as it gets dark.

Jacobus Cornelius Kapteyn

David M. Harland

One hundred years ago, on 18 June 1922, J. C. Kapteyn, one of the most distinguished astronomers of his generation, died in Amsterdam at the age of 71, possibly of a cancer that affected his white blood cells.

-oOo-

Jacobus Cornelius Kapteyn painted by Jan Pieter Veth in 1918 on the 40th anniversary of his appointment as a professor at the University of Groningen. On the wall is a portrait of the late David Gill. (Wikimedia Commons / Jan Pieter Veth)

Jacobus Cornelius Kapteyn was born on 19 January 1851 in the town of Barneveld in the Dutch province of Gelderland. He displayed early academic abilities and enrolled at the University of Utrecht at the age of 17, although he had achieved the entrance requirements the previous year, and studied mathematics and physics.

His astronomical career began immediately after graduation in 1875, when he was recruited as an observer by the Leiden Observatory. Three years later, he became the first Professor of Astronomy and Theoretical Mechanics at the University of Groningen. As the university possessed no research telescopes, Kapteyn concentrated on analysis. At that time, the primary focus of astronomy was mapping the stars on the celestial sphere.

The *Bonner Durchmusterung* (BD) star catalogue issued by F. W. A. Argelander in Germany between 1859 and 1862 was a massive project of visual observing. In 1871, Richard L. Maddox in England introduced a 'dry' gelatin emulsion for photography. On becoming director of the observatory at the Cape of Good Hope in South Africa in 1879, the first major observatory in the far southern hemisphere, Scotsman David Gill set out to photograph the southern sky. In 1885 it was agreed he would send the plates to Kapteyn, who would measure them using apparatus of his own design. The photography was completed in 1889. Although Kapteyn's task took twice as long as expected, the *Cape Photographic Durchmusterung* (CPD) was issued between 1896 and 1900.

After counting stars in selected areas of the sky, William Herschel inferred that the Milky Way is a slab-shaped system of stars with the Sun located close to the centre. Kapteyn decided to repeat this analysis, and launched an international effort to obtain the necessary plates. For each star, its brightness, spectral characteristics, radial velocity, proper motion, and so on, were determined. After another protracted effort of measurement and data reduction, his paper 'First Attempt at a Theory of the Arrangement and Motion of the Sidereal System', issued by the *Astrophysical Journal* in 1922 just a few weeks before his death, reported the Milky Way to be a lens-shaped stellar system 35,000 light years in diameter, with the Sun about 2,000 light years from the densely packed centre.

However, 'Kapteyn's Universe', as it became known, conflicted with a study reported several years earlier by Harlow Shapley at the Mount Wilson Observatory in California. His analysis of how globular clusters were distributed indicated the Milky Way system to be a disk of stars ten times larger than that suggested by Kapteyn. Its central point was in the direction of the constellation of Sagittarius, and the Sun was well to one side of the disk. For a reason that was not evident to Shapley, our view of the central region (and the far side of the disk) is obscured. The fact that all of the stars we observe are on our side of the disk explained

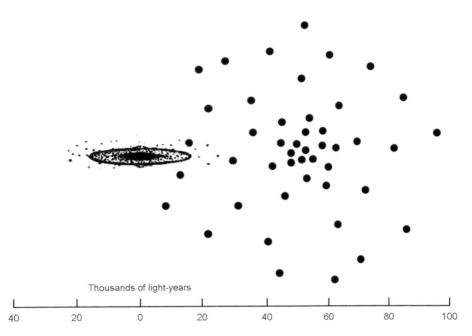

The distribution of globular clusters in space relative to 'Kapteyn's Universe'. (This is based on an original by J. H. Oort which appeared as Figure 1 in 'Lessons from the Milky Way' by P. C. van der Kruit, University of Groningen)

why, by counting stars, Herschel and Kapteyn had reckoned the Sun to be near the centre.[1]

In addition to the positions of stars, astronomers were eager to find out their distances. The parallax method was viable only for relatively nearby stars, so, in parallel with analysing plates for the CPD, Kapteyn had used the proper motions of stars to estimate their distances. At the September 1904 International Congress of Arts and Science in St. Louis, he reported that an analysis of the proper motions of the brighter stars in the solar neighbourhood indicated they were *not* moving randomly, as had been thought. There were two preferred directions, or 'apparent vertices'. After subtracting the motion of the Sun, which was imposing perspective

---

1. Shapley was correct in his determination of the shape of the Milky Way system, but he overestimated its size because at that time astronomers were unaware that light is attenuated by its passage through the gas and dust of the interstellar medium, which is thickest in the plane of the Milky Way.

Although Kapteyn never succeeded in establishing an observatory at Groningen, in 1896 he set up an Astronomical Laboratory. It became the Kapteyn Astronomical Institute upon his retirement in 1921. (Pieter C. van der Kruit/University of Groningen)

distortions, he gained the 'true vertices', which were in the plane of the Milky Way system and 180° apart, with one in Sagittarius. This 'streaming' was explained in 1925 by Bertil Lindblad at Uppsala University in Sweden, who realised that if Shapley was correct and the Sun was *not* close to the centre of the disk, and the stars were travelling around the centre in orbits that were only *approximately* circular, then at any given time roughly half of them would be increasing their orbital radius beyond their mean distance and the others would be reducing their radius, thereby giving the *illusion* of two streams of stars that were travelling in opposite directions. Kapteyn's data was the first evidence that the Milky Way system, which at that time appeared to be the entire universe, is in a state of rotation.

March

New Moon: 2 March
Full Moon: 18 March

MERCURY passes three of the four gas giants this month, skimming past Saturn on 2 March (0.7° separation), Jupiter on 21 March (1.2°) and Neptune two days later (0.9°). All of these planets are located in the east at dawn, with Mercury finishing what is the best morning apparition this year for southern hemisphere observers. It continues to brighten ahead of superior conjunction early next month, ending March at a dazzling magnitude −1.8, but look for it early in the month as it is getting lower to the horizon with each passing morning.

VENUS and Mars come together again this month, with the two planets separated in the dawn sky by 4.5° on the sixth day of March. Venus attains greatest elongation west (46.6°) on 20 March and theoretically reaches dichotomy the following day. It then has a close encounter with Saturn on 28 March. The morning star is best seen from tropical and southern latitudes. In a telescope, the bright planet appears as a waxing crescent until around the time of dichotomy (theoretical and apparent dichotomy often differ by several days) and a waxing gibbous disk afterwards; at the same time the apparent radius of the planet slowly diminishes as Venus leaves Earth behind.

EARTH reaches its first equinox of the year on 20 March, ushering in spring to the northern hemisphere and introducing autumn to the southern hemisphere. Earlier in the month, on 7 March, the Moon once again occults Uranus.

MARS departs Sagittarius for Capricornus on 5 March, the day before its second encounter of the year with Venus. Both planets are visible in the morning sky but Venus is much brighter than first-magnitude Mars. Early risers south of the equator will have the best opportunity to see the red planet pass just 0.2° north of the fourth-magnitude variable G-type star Iota (ι) Capricorni on the penultimate day of the month.

JUPITER is at conjunction on 5 March and may still be too close to the Sun ten days later for its extremely close encounter with the reddish fourth-magnitude star Phi (φ) Aquarii to be visible. It has a morning appointment with Mercury on 21 March when the two planets are just 1.2° apart. Southern hemisphere observers have the best views of this planet late in the month just before sunrise.

SATURN is approached by both inferior planets this month, with Mercury (magnitude −0.1) appearing 0.7° south of the ringed planet on 2 March and Venus (magnitude −4.4) passing 2.1° north of first-magnitude Saturn on 28 March. All of these objects inhabit the morning sky, with Saturn well-placed in Capricornus for planet watchers in southern latitudes. However, Saturn remains mired in morning twilight as seen from more northerly vantage points.

URANUS is again occulted by our satellite this month, with the faint planet disappearing behind the disk of the thin waxing crescent Moon on 7 March. Residents of New Zealand and nearby south-Pacific islands will have a chance to observe this event just after sunset. The green ice giant sets mid-evening so look for it as soon as the skies are properly dark.

NEPTUNE is at conjunction on 13 March and is too close to the Sun to observe this month. Its close approach to Mercury ten days later is unlikely to be visible in the dawn sky.

A True Pioneer of Planetary Exploration

Neil Haggath

50 years ago, NASA launched a small space probe, which would travel further into the Solar System than any before. Pioneer 10 was the first probe sent to Jupiter, and therefore the first to cross the Asteroid Belt. It was also the first to achieve solar escape velocity, meaning it will leave the Solar System and enter interstellar space.

Pioneer 10 was launched on 3 March 1972, by an Atlas-Centaur rocket, with an added third stage which had been specially developed for the purpose. It was accelerated to 51,670 kilometres per hour, faster than any previous spacecraft; it crossed the Moon's orbit after only 11 hours. The probe's mass is a mere 258 kg, and its main body is only 1.4 metres across, but its largest component is its high gain antenna, with a 2.74-metre parabolic reflector, with which to communicate with Earth from deep space. It carries 11 instruments, including a camera, infrared and ultraviolet detectors and a magnetometer.

As solar panels are not very efficient at the distance of Jupiter, it was the first spacecraft to be powered by radioisotope thermoelectric generators, which use the

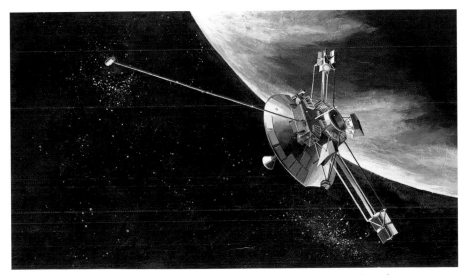

Artist's impression of Pioneer 10 passing Jupiter. (NASA/Rick Guidice/Wikimedia Commons)

heat released by the radioactive decay of plutonium-238. These are mounted on two three-metre booms, to prevent their radiation interfering with the instruments.

Pioneer 10 entered the Jovian system on 8 November 1973, when it crossed the orbit of Sinope, the outermost of Jupiter's known satellites. It made its closest approach to the giant planet on 5 December, passing at a distance of 130,000 kilometres. During the encounter, which formally ended on 2 January 1974, it transmitted over 500 images, and studied Jupiter's atmosphere and magnetic field. Precise tracking of its trajectory refined the measurement of the planet's mass, and infrared measurements showed that it emits more heat than it receives from the Sun. It was also discovered that Io, the innermost of the Galilean satellites, has an ionosphere, and orbits within an extended cloud of hydrogen.

Pioneer 10 was followed by the near-identical Pioneer 11, launched on 6 April 1973. When the latter flew by Jupiter in December 1974, it passed much closer than its predecessor, and used the planet's gravity to accelerate it onto a trajectory which would take it on to Saturn, which it passed in September 1979. (It was not the first probe to use a gravitational "slingshot"; Mariner 10 did so ten months earlier, using Venus' gravity to send it on to Mercury.)

A few years later came the two far larger and more sophisticated Voyager probes, launched in 1977. Both flew by Jupiter and Saturn, with Voyager 2 using further gravity assists to travel on to Uranus and Neptune. While the Voyagers made vastly greater contributions to our knowledge of the outer planets, the two tiny Pioneers had paved the way.

Pioneers 10 and 11 and the Voyagers are all on trajectories which will take them out of the Solar System into interstellar space. When they did so is a matter of definition, but is usually defined as crossing the heliopause, the boundary of the heliosphere, the Sun's huge bubble-like region of influence. The distance of the heliopause varies with solar activity, but averages about 123 astronomical units (au) from the Sun. The Voyagers, which are travelling faster and have "overtaken" the Pioneers, have already done so, in 2012 and 2018; when they did so could be determined, as they were still transmitting data on plasma density and magnetic fields.

Pioneer 10's mission formally ended on 31 March 1997, when it was 67 au from the Sun, but telemetry data was still occasionally received from it until April 2002, when it was about 80 au away. At the time of writing in early 2020, it is 125 au from the Sun, and is therefore crossing the heliopause almost as I write!

The probability of either of the Pioneers ever being intercepted by an alien civilisation is obviously miniscule but they each carry a "message from Earth", engraved on a gold anodised aluminium plaque mounted on the spacecraft. The message was devised by Carl Sagan and Frank Drake, and drawn by Sagan's then wife, the artist Linda Salzman Sagan.

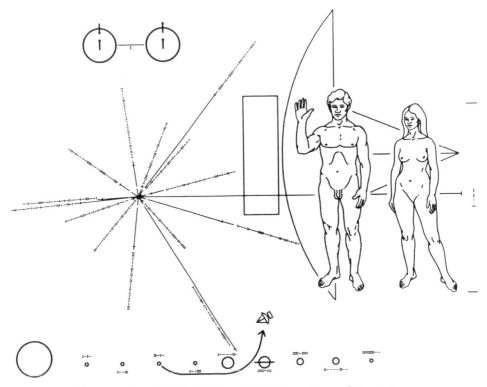

Pioneer 10's "message from Earth". (NASA) or (NASA/Ames Research Center)

Naturally, it shows what we look like! The human figures are shown against the spacecraft itself, to show our size. Equally obvious to us, but probably not to an alien, is the representation of our Solar System, showing the probe's path.

At upper left is a representation of the hyperfine transition of hydrogen, which is used to establish a system of units; the wavelength of the radiation emitted by that transition, 21.1 cm, is used as a unit of length, and the reciprocal of its frequency as a unit of time. The radiating lines indicate where the probe came from; they represent the relative directions and distances from the Sun of 14 known pulsars, with their periods in the established time units. Whether any intelligent alien would be able to decipher the message is anyone's guess!

The message caused considerable controversy. Significant numbers of people objected to the humans being shown naked, and protested about NASA "sending pornography into space"! Several American newspapers reproduced the message, but with the anatomical details removed. Laugh or cry as you see fit!

April

New Moon: 1 April
Full Moon: 16 April
New Moon: 30 April

MERCURY reaches superior conjunction on the second day of April and moves back into the evening sky. This apparition, which lasts until late May, is the best western appearance of the tiny planet for observers in northern temperate latitudes. It is also the least favourable evening apparition for those in the southern hemisphere. As is always the case with evening appearances of Mercury, the tiny planet begins very bright and then dims as it heads from superior to inferior conjunction, the exact opposite behaviour to Venus. The range of magnitudes this month is −1.8 to +0.6. On 9 April, Mercury passes through its ascending node and four days later reaches perihelion for the second time this year. After visiting three of the four gas giants last month, Mercury rounds out the list with Uranus on 18 April when the two planets are 1.9° apart. Greatest elongation east (20.6°) takes place on the penultimate day of the month, coinciding with a close approach to the open star cluster M45, the Pleiades.

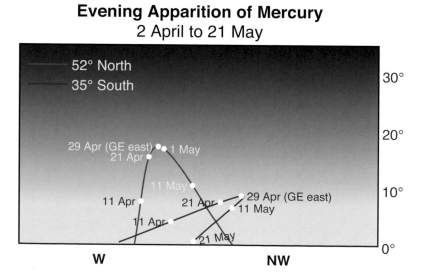

Evening Apparition of Mercury
2 April to 21 May

VENUS passes through its descending node on 10 April, passing north to south across the plane of the ecliptic. When viewed through a telescope, the morning star is diminishing in apparent radius (21.9 to 16.8 arc-seconds over the course of the month) even as the phase increases from 55% to 68%. However, the smaller size wins out over the increased phase and Venus's magnitude decreases slightly from −4.3 at the beginning of April to −4.1 at the end. Venus is high above the eastern horizon for observers in equatorial and southern regions; these early risers will get the best views of Venus passing 0.01° south of Neptune on 27 April and 0.2° south of Jupiter three days later.

EARTH experiences a partial solar eclipse on the last day of month, details of which are given in *Eclipses in 2022*. Another lunar occultation of Uranus takes place on the third and the Lyrid meteor shower occurs during the latter half of the month; for more information, see *Meteor Showers in 2022*.

MARS makes a close approach to Saturn on 5 April, with the two bodies closing to a distance of 0.3°. Mars is just the fainter of the two objects, shining at magnitude +1.1 compared to Saturn's +0.9. Mars leaves Capricornus for Aquarius six days later on 11 April. The red planet remains in the morning sky, best seen from southern latitudes where it rises around four hours before sunrise. For those in northern temperate latitudes, however, dawn is well underway before Mars appears in the east.

JUPITER encounters Neptune on 12 April, gliding just 0.1° north of its distant neighbour in Aquarius. Two days later it crosses the border into neighbouring Pisces and on the last day of the month, makes a brilliant pairing with the morning star. Venus is the brighter object at magnitude −4.1 with Jupiter some two magnitudes fainter. Early risers in the southern hemisphere will have the best views, with observers in northern temperate latitudes needing a low eastern horizon.

SATURN is overtaken by Mars on the fifth day of the month with the ringed planet the slightly brighter object. Found in Capricornus, it is most easily viewed from the southern hemisphere where it is well aloft before dawn brightens the sky.

URANUS participates in a lunar occultation event on 3 April; it is only visible from the south- and mid-Atlantic and parts of Africa (Sierra Leone, Liberia, Côte d'Ivoire and Ghana). Uranus is currently found in the west after sunset and disappears below the horizon during twilight hours. This will make it difficult to observe its close passage by Omicron (o) Arietis (a B-type main sequence star which is of a similar apparent brightness to Uranus) on 14 April and the Mercury fly-by four days later. Solar conjunction is early next month so look for Uranus at the beginning of April.

NEPTUNE appears in the morning skies this month. It is now past conjunction but still deep in dawn-lit skies when viewed from northern temperate latitudes. However, early risers in the southern hemisphere have a much better chance of seeing the blue planet when it encounters Jupiter on 12 April. These two gas giants last met in 2009 when they underwent a triple conjunction. This year they meet only once but approach to within 0.1°. The next such Jupiter–Neptune conjunction will take place in 2035 but even nearer to the Sun. Neptune also encounters brilliant Venus on 27 April when the two planets are less than 0.01° apart in the sky. A telescope will be necessary to spot Neptune (magnitude +7.9, apparent diameter 2.4 arc-seconds) next to the morning star (magnitude −4.1, apparent diameter 17.2 arc-seconds).

Biela's Comet
A Tale of Two Parts

Neil Norman

Comets are erratic and volatile creatures at the best of times. They can act as predicted and go through the motions. Alternatively they can go into sudden outburst, making them hundreds of times brighter in a short period of time, or literally disintegrate completely before our eyes. They are much like people; each has its own characteristics.

Enter the object that came to be known as Biela's Comet, and which was first discovered on 8 March 1772, when French astronomer Jacques Leibax Montaigne spotted it as a magnitude 6.5 object in Puppis. The comet was just 0.38 au from Earth and past perihelion; it could only be followed for 29 days, which did not allow a reasonable orbit to be calculated.

The second discovery was made by the great comet discoverer Jean-Louis Pons on 10 November 1805. The comet was then just visible to the naked eye at magnitude 4.5 in the constellation of Andromeda and some 0.29 au from Earth. The comet displayed a weak coma, doubtless due to it having made frequent perihelion passes of the Sun, resulting in its volatile ices being sublimated away to virtually nothing. Consequently, although a magnitude of 4.5 sounds relatively bright, it is in fact a hard object to see as the coma disperses the light of the Sun over a larger surface area of the coma.

The third discovery was made by German amateur astronomer Wilhelm von Biela on 27 February 1826, when the comet was found as a magnitude 8.5

Wilhelm von Biela. (John McCue)

object in Aries. It was followed for a total of 72 days, giving ample time for a reliable orbit to be determined. Comparisons were made with the previous apparitions of 1772 and 1805, and both Biela and French astronomer Jean-Félix Adolphe Gambart wrote independent letters on 22 March 1846, each providing parabolic orbits and statements that the orbital paths were very similar to the comets seen in 1772 and 1805. The letters were published in the same issue of *Astronomische Nachrichten*. Heinrich Wilhelm Matthias Olbers independently came to the same conclusion, and sent a letter a few days later. Danish astronomer Thomas Clausen and Gambart independently calculated short period orbits of 6.67 and 6.57 years respectively. Gambart added the observations of 1805 to his calculations, and in May determined the period as 6.74 years, which is very close to the actual period of 6.72 years.

The comet was recovered by John Frederick Herschel on 24 September 1832 and was followed through until 4 January 1833, when it was last reported by Thomas Henderson from the Royal Observatory at the Cape of Good Hope in South Africa. No tail was reported and the coma had a visual diameter of no more than three arc minutes.

Although the 1839 apparition was missed due to unfavourable viewing circumstances, the comet was recovered by Italian astronomer Francesco de Vico on 26 November 1845. It was fainter than expected but was to brighten steadily during December, and in mid-January 1846 observers were in for a surprise. The comet had split into two pieces, both of which were observed until the end of

Biela's Comet in February 1846 as depicted by Austrian astronomer Edmund Weiss in his Bilderatlas der Sternenwelt. Fragment A is clearly seen as the most prominent and active. (Wikimedia Commons / Edmund Weiss)

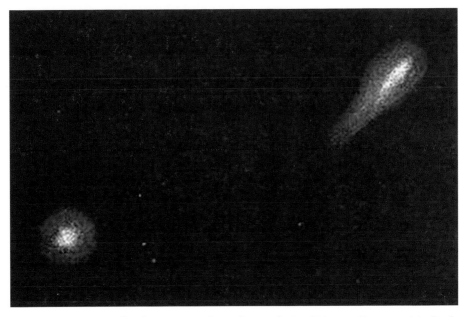

The two components of Biela's Comet as drawn by Angelo Secchi in 1852. Fragment A is clearly still the more prominent but is slowly fading when compared to the 1846 apparition. Illustration taken from *The World of Comets* by French science writer Amédée Guillemin. (Wikimedia Commons / Angelo Secchi / Amédée Guillemin)

March when one of the nuclei (the fainter one) was lost from sight. The brightest fragment was followed until 27 April.

The return of 1852 was eagerly awaited, and many questions needed to be answered, the most obvious being whether two comets would return? Or one comet? Or none? The answer was to come on 26 August of that year when Italian astronomer Angelo Secchi recovered the comet. However, only one of the fragments was seen until 15 September when the fainter section was picked up. This was not a favourable return, and the components were followed for only another 14 days, until 29 September when they were lost from view.

The apparition of 1859 was also unfavourable, with astronomers having no chance of locating the comet, so they had to wait until the next favourable return in 1865/66 to see just how much, if anything, of the pair had remained. However, search as they might, astronomers failed to locate the returning comet. This led some to predict that it had continued to break apart and that nothing was left, a bold prediction for the time as cometary science had not seen anything like this

This image from *Le Ciel* by Guillemin depicts the Andromedid (or Biela) meteor shower observed over France on the night of 27 November 1872. (Wikimedia Commons / Amédée Guillemin)

before, at least to such an extent. The return of 1872 was also favourable, and was considered as presenting the last good opportunity of a recovery. However, the year moved on and nothing was detected until the night of 27 November, when a meteor shower of epic proportions was seen.

So what had become of the comet? Was this the end of the story? On the contrary, this was only the beginning of a new branch of cometary science, as we will see when the second part of this article appears in the *Yearbook of Astronomy 2023*.

May

Full Moon: 16 May
New Moon: 30 May

MERCURY is becoming fainter with every passing day, starting the month at magnitude +0.6 before dimming to sixth magnitude around the time of inferior conjunction on 21 May and then recovering slightly to +2.8. This evening apparition favours northern observers but look for the tiny planet early in the month whilst it is still relatively bright and high in the west. The very young crescent Moon passes 1.8° south of Mercury on the second of May. Direct motion ends on 11 May and Mercury passes through its descending node six days later; it will be a morning sky object by the time it reaches aphelion on 27 May.

VENUS is the morning star, high in the east for those in southern latitudes who also rise ahead of the Sun but always quite low to the horizon when viewed from northern temperate regions. Our sister world reaches aphelion on 15 May and on 27 May it is occulted by the waning crescent Moon. This event begins around 00:30 UT and is visible from the southern part of Madagascar. The nearby islands of Mauritius and Réunion may also catch a glimpse of this occultation before sunrise. Venus loses only a tenth of a magnitude in brightness this month, ending May at a still brilliant −4.0. A telescope shows an increasingly small disk which is becoming more full with each passing morning.

EARTH undergoes a total lunar eclipse on 16 May; more information may be found in the article *Eclipses in 2022*. Other Moon-related events include three lunar occultations. Uranus is the first, disappearing behind the waxing crescent Moon on 1 May but as this is only one day past New Moon, this event will not be observable. Venus is occulted on 27 May and the second occultation of Uranus takes place the following day. The Eta Aquariids peak early in the month and should not be unduly inconvenienced by the waxing crescent Moon. The otherwise obscure Tau Herculid meteor shower may exhibit substantially increased activity from the end of the month with the New Moon providing perfect viewing conditions. See *Meteor Showers in 2022* for more details.

MARS appears half a degree south of Neptune on 18 May; although Mars is a first-magnitude celestial object, a telescope will be necessary to spot nearby eighth-magnitude Neptune. The following day, on 19 May, Mars crosses the border from Aquarius to Pisces. The waning crescent Moon pays a visit to the red planet on 24 May, passing 2.8° south of Mars. Five days later Mars encounters the much brighter Jupiter. Mars continues to linger in the morning sky, best seen from the southern hemisphere where it rises several hours ahead of the Sun. The disk of Mars is less than 90% illuminated this month and looks distinctly gibbous in a telescope.

JUPITER parts company with Venus but is overtaken by Mars on 29 May when the red planet is just 0.6° north of its giant neighbour. Those watching from southern latitudes continue to have the best views of Jupiter, slowly brightening from magnitude −2.1 to −2.2 in the sprawling constellation of Pisces.

SATURN continues to move across Capricornus in May although its progress is slowing with retrograde motion about to begin next month. The planet shines at magnitude +0.9 and reaches west quadrature on 15 May. Fifteen days later it attains its maximum northerly declination of the year. Saturn appears in the morning sky, rising around midnight for southern hemisphere observers and an hour or two later for those in the north.

URANUS is occulted by the Moon on the first day of the month but is at solar conjunction four days later so the green planet is lost in the glare of the Sun for all of early May. A second lunar occultation takes place on 28 May, when the very old crescent Moon is only 0.3° away from the green ice giant. However, only those floating off the Pacific coast of South America will have a chance to observe this event just before sunrise.

NEPTUNE moves into the constellation of Pisces where it will remain for the next two and a half months. It is visible in the morning sky and is best seen from the southern hemisphere. It is a difficult object for observers in northern temperate latitudes where it is only slowly emerging from morning twilight. It is found 0.5° north of the red planet on 18 May but planet watchers will need optical aids to see the eighth-magnitude Neptune near the first-magnitude Mars.

Maximilian Hell
A Legacy in Transit

Richard Hill

Rudolf Maximilian Hell (spelled Höll) was born 15 May 1720 in Schemnitz (or *Selmecbánya*), Kingdom of Hungary, the third son of the second marriage of Matthias Cornelius Hell, a mining engineer who had total of 21 children! The area was historically alternately German, Hungarian and Slovak (leading to multiple names for the same geographic locality as the reader will see) and while he could understand Hungarian and Slovakian languages, his commonly spoken language as he was growing up was German. Even so, Hell identified as a Hungarian.

Maximilian joined the Jesuit order in 1738 right after graduating from secondary school and made his novitiate (training in preparation to taking the "vows" in the church) in Trenčin. In 1745 he was invited to be an assistant to astronomer Josef Franz, then Director of the Jesuit observatory in Vienna, where he studied theology and helped in the founding of a museum in experimental physics.

Hell wrote many papers and became a well-known astronomy popularizer, his lectures often attracting large audiences. He was sent to be a teacher of humanities at the secondary school in Levoča (Slovakia) founding an observatory at Trnava at that time. Maximilian was subsequently ordained as a priest in 1752 and received his doctorate in mathematics, his first assignment being in Kolozsvár (Klausenburg) at the time

Portrait of Maximillian Hell at Vardø (artist unknown). Note the heavy clothing and the portion of one of his instruments that he is touching. Through the window you can see his quarters with the transit observatory. (Wikimedia Commons)

part of Hungary but today known as Cluj-Napoca, Romania. Here he was charged with the construction of a new college that included an astronomical observatory. He was also the instructor for mathematics, physics and history and published texts on these subjects. Clearly a man of extraordinary energy!

In 1755, Hell's lifelong career was set in place when Maria Theresa, Archduchess of Austria and Queen of Hungary and Bohemia (of the reigning Hapsburgs), named him as her "court astronomer" commissioning him to organize a new large observatory in Vienna, along with a general reform of the University of Vienna that would give it European prominence. Maximilian was appointed as the lifelong director, with an annual income of 300 golden crowns. The initial equipment was that used by the late Imperial Mathematician and geodetic surveyor, Johann Jakob Marinoni. His house and personal observatory were on the outskirts of Vienna a mile or so from the University, which in those days was a formidable distance. So the instruments were salvaged and installed in a tower four stories tall, over the newly completed hall of the university in Vienna. Here he quickly began making regular observations and establishing valuable connections with the European astronomical community and published *Ephemerides astronomicae ad meridianum Vindobonemsem* (Ephemerides for the Meridian of Vienna) only two years after being appointed! The only other such tables were published by the well-established Paris Observatory so this was a bold move and gave the Vienna Observatory some notoriety. Hell's Ephemerides laid out the theoretical predictions of the transits of Venus for 1761 and 1769 (which will become important later on) and how to derive the Sun's parallax from observations thus establishing a value for the astronomical unit. There were also tables for Sun, Moon and planets and explanations on how to observe them, discussions on meteorological observational methods (including his explanation of the aurora borealis) and a geodetic survey with latitudes and longitudes of places in northern Europe and the exact mapping of the empire.

He published this work each year up to and including 1768, missing four years while he was off on an observing expedition for the 1769 transit of Venus, and then picked it up again from 1772 to 1792, the year of his death. This work was so important that Franz de Paula Triesnecker, who had been appointed assistant director of the Vienna Observatory in 1782 (succeeding Hell as director in 1792) went on to publish a further 14 volumes of this work.

This new observatory and its ephemeris undoubtedly had some practical significance for the Archduchess since the Seven Years' War began in 1756 and Vienna itself was under threat of invasion almost at once. The ephemeris, with its improved positions of cities in Europe and concomitant improvement in mapping of the continent would aid in the movement of troops and planning of battles. This could explain the seemingly hurried first ephemeris publication only two years after the founding of the observatory.

The transit of Venus which occurred on 6 June 1761 was an important astronomical event for the reasons mentioned above. In Vienna the whole event could be seen and was observed by Hell and César-François Cassini de Thury (the grandson of the famous Italian astronomer Giovanni Domenico Cassini). Because it was only partially observable from Paris, Cassini planned to travel only as far east as Vienna to observe the event, a restriction undoubtedly forced on him by the Seven Years' War which was then in full vigour north and east of Vienna.

Their results were among those that were too poor for the primary scientific goal of transit observations, the determination of the astronomical unit. In fact only the observations carried out by English astronomers Charles Mason and Jeremiah Dixon (of Mason-Dixon fame in U.S. geographic history) from the Cape of Good Hope were usable, and even they carried some significant errors. Therefore astronomers around the world were dedicated to the goal of improving instrumentation and techniques for the transit in 1769.

In 1767 Maximilian received an invitation from King Christian VII of the combined kingdom of Denmark and Norway to observe the transit of Venus on 3 June 1769 from the small town of Vardø in the province of Finnmark, the most north eastern settlement in Norway. Situated nearly 4° above the Arctic Circle, this site had the advantage of 24 hours of daylight. The offer came as a surprise as Hell had never had any direct contact with any astronomical institute in Denmark.

An extract from the northernmost section of the Map of Scandinavia, Norway, Sweden, Denmark, Finland and the Baltics drawn up by the German geographer and cartographer Johann Baptist Homann, showing the exact location (in 1730 terms) of Vardø, depicted here as Wardhus but on the title page of his published observations as WARDOËHUSII). (Wikipedia/ Geographicus Rare Antique Maps)

However, it may have been desirable to leave Vienna at that time for while the Seven Years' War had ended, there was now a suppression of Jesuits taking place in Europe. For Hell it would culminate with Jesuits being expelled from Austria in 1770 (though he seems to have avoided that). King Christian was a Protestant, and Maximilian would have his protection while carrying out this expedition. The invitation was accepted and he took a former student and fellow Jesuit priest János Sajnovics with him.

They took a very circuitous route up through Norway before putting out to sea at Trondheim and sailing north around the northern end of Scandinavia and then to Vardø. One of their first tasks was to accurately determine the latitude and longitude of their observing site which was probably valuable information to the government since the province of Finnmark had been aquired only relatively recently. Maps of that period show how little was known of the area.

Hell saw the transit as only one part of a larger scientific expedition, Hell and his companion also making magnetic observations which showed diurnal variation in the magnetic declination as well as some magnetic storms. In fact, Hell and his team stayed in Vardø for over two years, spending most of their time collecting other scientific data for an encyclopedia concerning on arctic regions they planned to publish. This was to contain studies in biology, meteorology, oceanography, zoology, geography, natural history and linguistic analysis. Sajnovics was studied in linguistics, and helped to explore the already widely discussed but not well understood similarities between the language of the Sami, Finns and the Hungarians. Sadly, nothing came of Hell's encyclopedic tome, due to the ongoing Suppression of the Jesuits throughout Catholic Europe culminating in 1773 when Pope Clement XIV recognized the suppression as a "fait accompli" in his brief, *Dominus ac Redemptor* essentially giving it his blessing. Nevertheless they did publish *Demonstratio idioma Ungarorum et Lapponum idem esse* (Copenhagen, 1770) on the language study through their Protestant patron.

Meanwhile the astronomers who had stayed home were anxious to get his observations, since Hell's were expected to be the most reliable of all the scientists involved. However, Hell felt his first duty was to report to his Danish sponsor. The Jesuits being specifically expelled from Austria in 1770, likely had some bearing on his lack of haste. This upset the French Academy which prompted Jérôme Lalande to accuse Hell of having falsified his data to better agree with observations of other observers. This was a devastating accusation but not surprising in that the French were the main agitators against the Jesuits at the time. The observations and detailed analysis of Hell and Sajnovics were published in 1770, the same year as their linguistic study, by the Royal Danish Academy of Sciences and Letters (Protestant) as *Observatio transitus Veneris ante discum Solis die 3. Junii anno 1769*

(Copenhagen, 1770), only one year after the event and which at that time was not unreasonable! So it would seem that there may be more politics than science behind this.

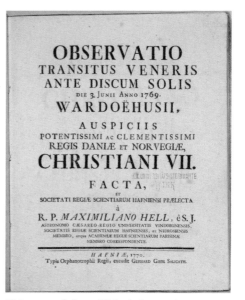

The Jesuits were not restored until 1814 (by Pope Pius VII) when the Catholic Church was seeking more support to rebuild and restart their schools after both the French Revolution and the Napoleonic Wars. Even so, the defamation of the Jesuits, as an attitude if not stated policy, lasted well into the nineteenth century when in 1823 Johann Franz Encke furthered the division in his review of the Venus transits in which he rejected Hell's work out of hand, with no reason given. Later, Joseph Johann von Littrow, the Director of the Vienna Observatory, misinterpreted

Title page of the published observations by the Hell expedition (1770). (Wikimedia Commons/Prof. Franz Kerschbaum, Institut für Astronomie, Universität Wien)

the fragmentary manuscripts he had seen of Hell's work and claimed he had seen erasures in them, announcing that he had thus found direct evidence of falsified observations. This was very damning as it came from an astronomer at the same institution, and it further ruined the reputation of Hell for over fifty years.

Hell's reputation was restored in 1883 by the American astronomer, Simon Newcomb who studied his observations cn toto preserved at the Vienna Observatory, and showed that in fact Littrow's inferences were wrong. This was verified when Hell's observations, properly included with those of others from 1769, showed the Sun's parallax much closer to the correct value than without them!

Even with all Maximilian had to deal with, he was still greatly admired throughout Europe for his scientific abilities and his kind and friendly demeanor. He was elected to numerous academies in Europe even after the accusations, and offered honorary pensions beyond his salaries, though he refused them. Today we should remember Maximilian Hell as a well-liked and well respected astronomer and overall renaissance scientist who, through politics, was denied his rightful place among the greatest astronomers in history, though it is hoped that this legacy is in transit!

June

Full Moon: 14 June
New Moon: 29 June

MERCURY rejoins Venus at dawn in what is another decent morning apparition for those in equatorial and southern latitudes. It brightens steadily, from magnitude +2.8 to −0.7, as it departs from inferior conjunction. Direct motion returns on the third day of the month and greatest elongation west, measured at 23.2°, occurs on 16 June. The closest planet to the Sun is best viewed during the latter half of the month when it is highest in the sky.

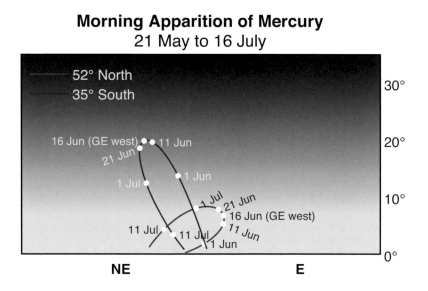

Morning Apparition of Mercury
21 May to 16 July

VENUS has been losing altitude above the eastern horizon for the past two months as seen from the southern hemisphere and June is no exception. However, the view from northern temperate latitudes is improving as Venus climbs slightly higher in the sky. It remains very bright (decreasing in magnitude from −4.0 to −3.9) but is getting ever smaller in the telescopic eyepiece. The increase in phase cannot compensate for the increase in distance from Earth as Venus heads toward the far side of the Sun. Venus moves past Uranus on 11 June and is itself overtaken by the waning crescent Moon on 27 June but no lunar occultation takes place this month.

EARTH arrives at solstice on 21 June; summer begins in the northern hemisphere and winter takes hold in the south. The Moon occults two planets this month, Mars on 22 June and Uranus two days later. The farthest apogee of the year takes place on 29 June. The Tau Herculids peak early this month; see *Meteor Showers in 2022* for more information.

MARS spends most of June in Pisces, dipping briefly into Cetus early in the month. Mars reaches perihelion on 21 June and is occulted by the waning crescent Moon the following day. However, only the denizens of Antarctica will have the opportunity to see this event. Mars is becoming easier to see from northern temperate latitudes where it rises just after midnight. A telescope reveals a rusty red gibbous orb.

JUPITER rises around midnight, making it a morning sky object. The Last Quarter Moon passes just 2.7° south of the magnitude −2.3 planet on 21 June. Four days later, Jupiter forsakes the zodiac and enters the constellation of Cetus where it will remain until September. West quadrature takes place on the penultimate day of the month.

SATURN enters into retrograde motion this month, reversing its course across Capricornus. The rings close to their minimum aspect for the year, 12.3°, on the first day of the month but the planet continues to slowly brighten as it approaches opposition in August, beginning June at magnitude +0.8 and ending at +0.6. The apparent diameter of its disk is also increasing as the planet nears Earth, starting at 17.4 arc-seconds and growing to 18.2 arc-seconds by the end of the month.

URANUS is past conjunction and is now part of the morning sky line-up. It is lost in the dawn skies for early risers in northern temperate latitudes but observers in the southern hemisphere have a chance of spotting it in the east before first light. It is 1.5° south of the morning star on 11 June and is occulted by the waning crescent Moon on 25 June. This last event is visible before sunrise from western and northern Australia, East Timor, Indonesia and other nations bordering the South China Sea.

NEPTUNE reaches west quadrature – 90° from the Sun – on 16 June. Now located in Pisces, Neptune rises around midnight and is accessible to anyone in either hemisphere with a telescope. It reaches its maximum declination north for the year on 24 June and begins retrograde motion four days later.

Margherita Hack

Mary McIntyre

Margherita Hack was an Italian astrophysicist, who is probably best known for her work on stellar spectroscopy. However, she was a prominent figure within the astronomy community and her research covered a much wider range of astronomical subjects. Margherita collaborated extensively with astronomers around the world and was a fantastic science communicator.

She was born in Florence on 12 June 1922 and this year marks the one hundredth anniversary of her birth. She enrolled at university in 1940, initially to study literature, and attended one class, which she claimed was "an hour of pointless chat in which I was not interested", after which she switched to a physics course. Her first astronomy class was in her third year at university and she went on to write her thesis on Cepheid variables. She graduated with top grades in 1945, a year after marrying Aldo de Rosa.

Margherita began her career as a staff researcher at the Astronomical Observatory of Arcetri in Florence. In 1953 she worked for six months at the Institut d'Astrophysique in Paris, then from 1954 to 1964 at the Observatory of Brera in Milan. In 1955 she met the Russian-American astronomer Otto Struve (1897–1963) from the Astronomical Department of the University of California, and one of the most important collaborations of her career began. She travelled to California to work with Struve on some important theories relating to the star Epsilon Aurigae – theories proved several years later by data from the Copernicus satellite. She visited California again in 1958, when she and Struve began working on their book *Stellar Spectroscopy*. Although Struve died in 1963, Margherita went on to finish the book herself, which was eventually published in two volumes in 1969. *Stellar Spectroscopy* remains an important textbook in the study of the atmospheres of stars.

Margherita Hack. (Wikimedia Commons/ Cirone-Musi)

In 1964 she was appointed a full Professor of Astronomy at the University of Trieste, making her the first woman in Italy to become a full professor. She was also the first female director of an observatory after taking over the Astronomical Observatory of Trieste, a position she held until 1987. Additionally, she was the director of the Astronomy Department of Trieste University from 1985 to 1991 and again from 1994 to 1997. Within a few years of her appointment she had transformed the small and little-known observatory into a modern research centre which went on to gain international success and much collaboration from around the world.

Margherita was one of the first astronomers in Italy to understand the importance of space telescopes in astronomy and cosmology research. During the 1970s she worked on UV data from the Copernicus satellite in order to study stellar atmospheres and mass loss, this research going on to confirm her and Struve's earlier theories. Her first paper based on this research was published in the journal *Nature* entitled 'Copernicus Spectra of β Lyrae'.[1] In the 1980s she embarked on a campaign to make her colleagues aware of the research opportunities available when the International Ultraviolet Explorer (IUE) was launched. This resulted in such a large number of Italian researchers having proposals accepted that in her autobiography she jokingly says that some of her European colleagues referred to it as the Italian Ultraviolet Explorer!

Following her retirement in 1997, she still kept up to date with current research projects at Trieste and retained her office on site. Although she had stepped away from her direct involvement with research, she increased her outreach and education activities and brought science and astronomy to both the general public and to a younger audience. In 1978 she co-founded the bimonthly magazine *L'Astronomia* with the Italian astrophysicist and science journalist Corrado Lamberti, and in 2002 she and Lamberti co-directed the popular science and astronomy magazine *Le Stelle*.

In 2011 she co-hosted a TV program called "Big Bang – Travelling Through Space with Margherita Hack" which aired on the DeA Kids TV channel as well making numerous other television appearances. Between 1996 and 2013 she also wrote a column in an Italian newspaper and was very active in politics.

From a young age Margherita was very passionate about politics, civil rights and equality and held strong views on pseudoscience, superstition and religion. In her

---

1. 'Copernicus Spectra of β Lyrae', M. Hack, J. B. Hutchings, Y. Kondo, G. E. McCluskey, M. Plavec and R. S. Polidan, *Nature*, vol 249, issue 5457, pp. 534–536 (1974)

The headquarters of Trieste Astronomical Observatory. (Wikimedia Commons)

final year of secondary school she was suspended for twenty days because of her anti-fascist views, which would have meant she was unable to sit her final school exams – had war not broken out, leading to cancellation of all exams that year! She felt very strongly about the lack of proper funding for scientific research within her country.

Margherita died on 29 June 2013 aged 91. During her career she published over 400 research papers (which can be viewed in the NASA archive) as well as over 20 books on popular astronomy. She had been part of many committees and working groups, including ESA, NASA and IAU. She received numerous awards and honours, including the Gold Medal of the Italian Order of Merit (1998).

On her 90th birthday she was awarded the title of Dama di Gran Croce from President Giorgio Napolitano, the highest honour of the Italian Republic. Her strong personality, scientific vision, management skills and ability to communicate science to a wide audience made her a truly formidable character, and she left behind an incredible legacy. The minor planet 8558 Hack, discovered in August 1995, was named in her honour.

July

Full Moon: 13 July
New Moon: 28 July

MERCURY is a morning sky object for the first half of the month. It passes through its ascending node on 6 July and four days later is both at perihelion and its maximum northerly declination for the year. Superior conjunction occurs on 16 July, after which the planet moves into the evening sky. This evening apparition favours southern latitudes but is very disappointing for those living farther north. Mercury is brighter than zero magnitude all month.

VENUS reaches its maximum northerly declination this month but it is rapidly losing height above the eastern horizon as seen from southern and equatorial latitudes. The planet has an apparent visual magnitude of −3.9 which it will have for the rest of the year. Superior conjunction is still three months away but the telescope reveals a small (12.0 down to 10.8 arc-seconds over the course of the month) disk which is nearly full (86% up to 93%). Venus is over 1.5 au from Earth by the end of July.

EARTH reaches aphelion on 4 July, the point in its orbit farthest from the Sun. This year's distance is 1.017 au or 152,100,000 km. The closest perigee of the year occurs on 13 July, coinciding with Full Moon, making this particular Full Moon, at 2005 arc-seconds wide, the largest of 2022. See the Full Moon of January for comparison. The Delta Aquariid meteor shower at the end of the month benefits from moonless skies; see *Meteor Showers in 2022* for observing details.

MARS is found just 0.2° south of the fourth-magnitude binary star Omicron (o) Piscium (the brighter component of this system is a K-type giant called Torcular) on the second day of July. The red planet then leaves Pisces for Aries six days later on 8 July. The waning crescent Moon, just a day past Last Quarter, occults first-magnitude Mars on 21 July. Whereas last month's occultation was seen from Antarctica, this one is visible in the Arctic where the Sun never sets at this time of year.

JUPITER now rises just before midnight but is still primarily a morning sky object in Cetus. It continues to brighten as it approaches its September opposition, increasing three-tenths of a magnitude to end July at −2.7. The waning gibbous Moon passes 2.2° south of the gas giant on 19 July. Jupiter reaches its maximum declination north for the year on 25 July; four days later it enters into retrograde motion.

SATURN continues to brighten and grow larger in telescopes as it nears opposition (and closest approach to Earth) next month. Found in Capricornus, the ringed planet is well-placed for viewing from the southern hemisphere as it rises during mid-evening hours. Astronomers in north temperate latitudes have had to wait a little longer for the planet to appear in the east but it now rises before midnight even for these observers.

URANUS is eclipsed by the waning crescent Moon on 22 July in a morning event visible from Brazil and north western Africa. Appearing in Aries, the planet is now rising early enough to appear in reasonably dark skies in the northern hemisphere but it is best viewed from southern latitudes. Look for it in the east after midnight.

NEPTUNE now rises before midnight in the zodiacal constellation of Pisces. At eighth magnitude, it is too faint to be seen with the naked eye but is visible as a tiny blue orb in a small telescope.

Gravity Assists – Something for Nothing?

Peter Rea

I once heard that there is no such thing as a free lunch. Somebody usually wants something in return. So what have free lunches and planetary exploration got to do with each other? An explanation is required.

There are many rockets or launch vehicles currently in use. Some larger than others. Each will have a payload limit depending on the target orbit. In planetary exploration the intended orbit is usually heliocentric, meaning the shape is an ellipse with the Sun at one focus in accordance with Kepler's first law of planetary motion. Some spacecraft like Voyager 1 and 2 were launched on hyperbolic trajectories and were given solar system escape velocities meaning they are leaving the solar system, never to return.

When proposing a planetary mission, cost is one of many factors to consider. If the mission objectives can be met using one of the smaller launch vehicles, cost can be kept down and the proposal is more likely to be funded. Sometimes space agencies have big ambitions. NASA's Cassini mission to Saturn is one example. Fully fuelled and ready for launch the Cassini orbiter and the Titan atmospheric probe Huygens had a launch mass of 5.7 tonnes. Only problem was, the Titan 4 Centaur, the biggest launch vehicle in the NASA inventory at the time could not launch this mass directly to Saturn. But Cassini / Huygens made it all the way to Saturn for a highly successful mission. How can this be so? It picked up energy en-route to Saturn by flying past Venus twice, then the Earth and finally Jupiter. How does this work?

If you look at Figure 1 you will see a flyby of a stationary planet. Not possible of course but shown for explanation. As the planet is not moving it possesses no energy other than gravity and in this scenario the velocity of the spacecraft will increase as it approaches the planet reaching maximum velocity at the closest point to the planet. As the distance to the planet increases on the way out, gravity will tug on the spacecraft causing it to slow down. Any velocity gained by gravity is lost on the way out. The net result being that the velocity out is the same as the velocity in.

However, planets are not stationary and follow their orbital paths around the Sun with differing velocities depending on their distance from the Sun. Kepler's laws of planetary motion again. More importantly the planets possess angular momentum, and this is the key to gravity assists. In Figure 2 we can now see an extra component,

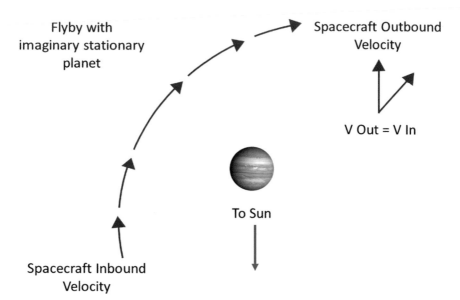

Figure 1: Gravity assist diagram 1
Planetary flyby with respect to a stationary planet. Velocity out is same as velocity in. (Peter Rea)

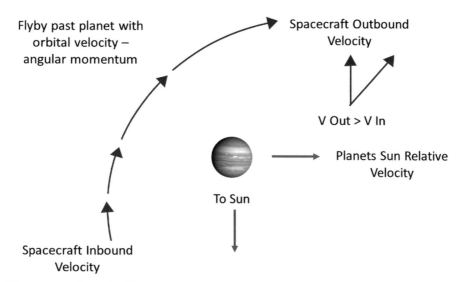

Figure 2: Gravity assist diagram 2
Planetary flyby with respect to a moving planet. Angular momentum is exchanged. Velocity out is greater than velocity in. (Peter Rea)

orbital velocity. As the spacecraft moves toward the planet the spacecraft's velocity will increase due to gravitational attraction. At closest approach angular momentum between the planet and the spacecraft is exchanged. If the flyby is from behind the planet then this exchange of angular momentum will result in an increase in orbital velocity of the spacecraft. If the flyby is in front of the planet then the exchange of angular momentum results in a decrease in the spacecraft orbital velocity. Both methods have been used. Missions wanting to go to the inner planets Venus and Mercury need to lose energy. Recall that as the Earth travels around the Sun its average orbital velocity is 110,000 kph. It varies slightly as Earth's orbit is an ellipse not a circle (Kepler again). To reach Mercury we need to reduce this velocity to enable the spacecraft to "fall" inward to the Sun. The first example of a gravity assist was Mariner 10 which used the gravity of Venus on 5 February 1974 to reshape the orbit to enable a flyby of Mercury on 29 March 1974.

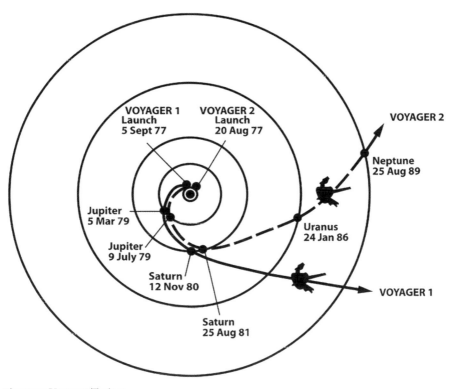

Figure 3: Voyager Trajectory
The trajectories of Voyagers 1 and 2 to the outer planets. Such an alignment occurs only once every 175 years. (NASA)

Missions to the outer planets need to gain energy rather than lose it. One of the best examples of this was the twin Voyager missions launched in 1977. Voyager 1 used the gravity of Jupiter to send it onto Saturn and then used Saturn to deflect the flight path of Voyager 1 downward to the large moon of Saturn called Titan. Voyager 2 went one better and reached all four of the gas giants. This alignment of the outer planets was discovered by the Jet Propulsion Laboratory engineer Gary Flandro. His work on spacecraft trajectories to the outer planets during the 1960s made the Voyager mission possible. This alignment of the outer planets occurs once every 175 years. The Voyagers had to be launched between 1977 and 1980 or the planets would move out of position. The Voyager project was a stunning example of gravity assists which has now become the norm in outer solar system exploration. Without it, planetary exploration would consist of a flying fuel tank with a few instruments on board and would require a huge launch vehicle and a very, very big budget. Other examples of gravity assists are the Galileo mission to Jupiter launched in 1989 which flew past Venus once and Earth twice to gain enough energy to reach Jupiter in 1995.

Gravity assists can help change the trajectory in another way. In Figure 4 we see the trajectory of the Ulysses spacecraft which launched in 1990. The objective of this mission was to fly over the north and south poles of the Sun, which is very costly

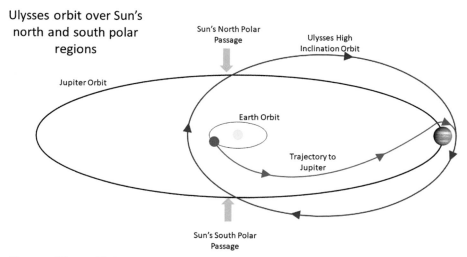

Figure 4: Ulysses Trajectory
The gravity of Jupiter was used to increase the orbital inclination of Ulysses to 80 degrees, allowing repeated passages over the Sun's north and south polar regions. The journey to Jupiter is shown in blue and the final Ulysses orbit in red. (Peter Rea)

in terms of energy required and practically impossible using conventional onboard propulsion. Jovian gravity to the rescue. On 8 February 1992, Ulysses flew over the northern part of Jupiter where the enormous gravitational field of Jupiter bent the trajectory southward increasing the inclination to the ecliptic by 80 degrees. This put it into a final orbit around the Sun that would take it past the Sun's north and south poles. The aphelion remained at the distance of Jupiter which it returned to every six years, although Jupiter was not there so the orbit remained the same. To achieve this manoeuvre without a Jupiter gravity assist would have required huge amounts of fuel and made the mission impossible as no launch vehicle existed that could have launched it.

So as we have just seen, gravity assists, or flybys can assist a spacecraft in the following ways.

1 – Change the direction of the trajectory
2 – Increase heliocentric velocity
3   Decrease heliocentric velocity
4 – Increase the inclination of the spacecrafts orbit

Back to the question of whether we get something for nothing. When a spacecraft passes a planet, angular momentum is exchanged. The spacecraft gains energy but also gives up some angular momentum. The huge difference in mass between the Galileo spacecraft (2,500 kg) and Venus (4.868×10^{24} kg) means that Venus looses very little energy. Long time friend and regular contributor to this publication Neil Haggath reminded me of the following fact. Someone at the Jet Propulsion Laboratory in California with way too much time on their hands calculated the changes in the velocities of Venus and Earth due to Galileo's gravity assists. Venus was slowed down by 25 millimetres per billion years, and Earth, due to the combined effect of the two flybys, by 150 millimetres per billion years. How do you go about working that out? No time to answer, just been invited to a free lunch.

August

Full Moon: 12 August
New Moon: 27 August

MERCURY soars into the evening sky for southern hemisphere astronomers in what is the best western appearance this year of this tiny planet. However, Mercury remains very low to the horizon as seen from northern temperate latitudes and may be inaccessible for many. Mercury passes 0.6° north of Regulus, the brightest star in the constellation of Leo, on 4 August. The closest planet to the Sun moves north to south across the plane of the ecliptic on 13 August; ten days later it is at aphelion. Greatest elongation east at 27.3° takes place on 27 August and is the largest one of the year. Mercury is dimming even as it gains altitude this month, beginning at magnitude −0.6 and ending at +0.4.

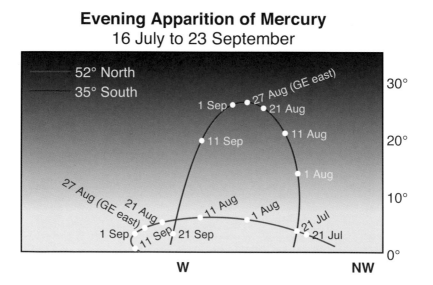

Evening Apparition of Mercury
16 July to 23 September

VENUS moves through its ascending node on the second day of the month. On 17 August, it is found just 0.9° south of the Beehive Cluster (M44) in the constellation of Cancer. Shining at magnitude −3.9, Venus is the morning star but it is getting much lower in the eastern skies at dawn. Observers in equatorial regions arguably have the best views of this bright planet this month.

Saturn

January to December 2022

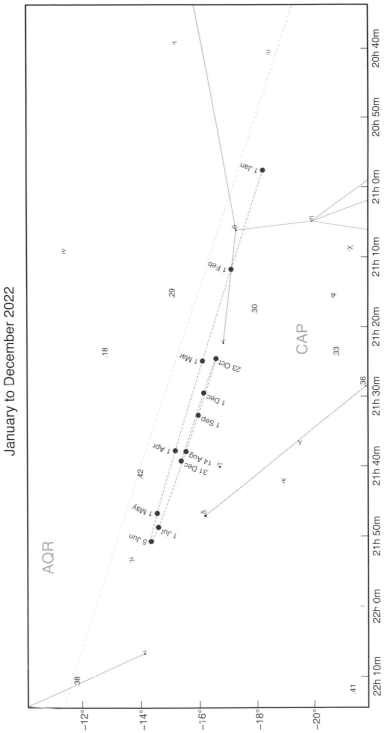

Background stars are shown to magnitude +5.5.

EARTH suffers through a poor apparition of the Perseids this month, with the Full Moon coinciding with peak activity; more details are in the article *Meteor Showers in 2022*. Uranus once again falls victim to the Moon, with the dim planet obscured by our satellite on 18 August.

MARS undergoes its last planetary conjunction of the year on the second of August when it moves past Uranus, gliding 1.3° south of the sixth-magnitude gas giant. One week later it departs Pisces and enters the constellation of Taurus where it remains for the rest of the year. Mars has been occulted by the Moon the past two months but on 19 August, the Last Quarter Moon gets no closer than 2.7° to the red planet. Mars reaches west quadrature on 27 August and presents a noticeably gibbous appearance to telescopic observers. Mars now rises before midnight for those in northern temperate latitudes but it remains strictly a morning sky object for astronomers south of the equator.

JUPITER and the waning gibbous Moon have a close encounter on 15 August, with the two celestial bodies less than 2° apart in the non-zodiacal constellation of Cetus. Jupiter continues to brighten, attaining magnitude −2.9 by the end of the month. The planet now rises in mid-evening for all observers.

SATURN is at opposition on 14 August, shining brightly at magnitude +0.3 in the constellation of Capricornus. It appears as a disk 18.8 arc-seconds wide (42.7 arc-seconds including the rings which are tilted at 13.9°) in the eyepiece of a telescope. Saturn is visible all night, rising at sunset and setting as the sun rises.

URANUS slows its progress across Aries this month as it enters into retrograde motion on 24 August. Prior to that, the sixth-magnitude planet appears 1.3° north of the much-brighter Mars on the second day of August and reaches west quadrature on the eleventh. One week later, on 18 August, it is occulted by the waning gibbous Moon; this event is visible from the northern Pacific, the far north eastern tip of Russia and the western reaches of Alaska just before sunrise. On 23 August, Uranus reaches its maximum declination north for the year.

NEPTUNE is in retrograde this month and backs out of Pisces, returning to Aquarius where it began the year and where it will remain for the rest of 2022. The eighth-magnitude planet rises in mid-evening and remains aloft until sunrise.

U.S. Mints Celestial Themes
Rare USA Coins Depicting Astronomy and Space

Carolyn Kennett

Coins have been used throughout history to represent the aspirations and prowess of the nations that they represent. Each country follows their own traditions, although the obverse of a coin is usually used to represent a head of a prominent person, in the UK it is used for the monarch's face, whereas the reverse can depict a variety of imagery from flora and fauna to famous people and concepts. Astronomy is just one subject which can get represented on coins and although not many examples come from the USA, there are three which we will explore here.

On 4 July 1776 the United States declared independence from Britain. This would lead to an eight year period of unrest which would mark the end of British rule. Right from the beginning a new country needed a new currency and early attempts to celebrate their autonomy included the creation of coins and the Continental Congress started to issue its own currency. In late 1776 a Silver Continental Dollar was struck with an astronomical design. The face of the coin depicts a sundial, rising above it and shining down upon it is the Sun. The accompanying words MIND YOUR BUSINESS and *Fugio*, (meaning *I fly*), were interpreted as a message for the British establishment. On the rear are thirteen interlocking rings representing the thirteen colonies (states). It has been said that the prominent figure Benjamin Franklin was the inspiration for the design but evidence shows that these were possibly minted in the United Kingdom and shipped to the new nation.

The 1776 Silver Continental Dollar. It is thought that Benjamin Franklin designed both sides of this coin, the obverse of which (seen here) depicts a sunrise over a sundial. (Wikimedia Commons/ Professional Coin Grading Service)

The reason why these coins went into disuse and were replaced is that the British forces were flooding the market with counterfeit versions. Benjamin Franklin, who was considered the inspiration for the design, said "The artists they employed performed so well that immense quantities of these counterfeits which issued from the British government in New York, were circulated among the inhabitants of all the states, before the fraud was detected. This operated significantly in depreciating the whole mass." Along with the depreciation of the currency due to unregulated production between the thirteen states this gave rise to the phase "not worth a continental". At the time many of these coins were struck in pewter, a rare few were made of silver or brass. These coins are now extremely rare and valuable as they symbolise America as an infant nation.

Between 1778 and 1789 The Articles of Confederation permitted the thirteen states to produce their own coins. In 1785 Vermont struck a coin which had the Sun rising over a mountaintop, which is covered in pine trees. Below this is a plough on the lush fields of verdant countryside.

Although there are only a handful of examples of astronomical themed coins, all American coins contain stars and rays and have done since the first coins were minted. When the War of Independence concluded in 1783, Americans still used the English reckoning of pounds, shillings and pence. American statesman Gouverneur Morris, the assistant superintendent of finance, proposed a decimal system, to replace the awkward division of the pound. Gouverneur Morris decided to mint copper coins using on the obverse imagery of the All-Seeing-Eye of God, surrounded by rays and stars. These pieces, known as Nova Constellatio (the first coins that were struck under the authority of the United States of America), were the first coins to bear American imagery and inscriptions. Once coins were regulated it was deemed that they would include Liberty, In God we Trust, and the four digits of the year on the obverse. The reverse has United States of America, *E Pluribus Unum* (the motto of the United States, meaning *out of many, one*) and the written value of the currency. There is little room for variety of other subjects.

A 1785 copper coin depicting the sunrise over a plough and the verdant fields of Vermont. This early American coin was struck for one year in just one state Vermont, post declaration. (Wikimedia Commons/Reuben Harmon)

Commemorative coins first appeared in 1892 and were mostly half dollars and depicted a wide range of subject from Spanish Queens to the Battle of Gettysburg. In the original run there were no astronomical themed commemorative coins and production ended in 1954 with a design of George Washington Carver and Booker T. Washington which was originally minted in 1951.

Then on 23 December 1981, thirty years after the last commemorative coin Congress approved a new design, something it had not done for thirty years. It would be a silver half dollar celebrating 250th anniversary of the birth of George Washington. It would take nearly another thirty years to get a space themed coin. In 2019 a design which celebrates space was finally approved. These coins were created to commemorate the 50th anniversary of the Apollo landings on the Moon. There are four coins in the set and they are already a collector's item.

The Apollo 11 50th Anniversary Commemorative Coin Program produced four coins. This is the obverse of the five-ounce silver dollar and shows a footprint made on the surface of the Moon. (Wikimedia Commons/US Mint)

This series will continue with a short article about coins which have travelled into space, telling us why they were sent and what were they used for. 'Giant Leaps for Small Change' will appear in the forthcoming *Yearbook of Astronomy 2023*.

September

Full Moon: 10 September
New Moon: 25 September

MERCURY is best observed from the southern hemisphere early in the month because it is both dimming and sinking back toward the western horizon. Mercury begins retrograde motion on 9 September which ushers in inferior conjunction on 23 September. It then rejoins Venus in the dawn sky, with the two heavenly bodies 3.2° apart three days later on 26 September.

VENUS reaches its second perihelion of the terrestrial year on 4 September, the previous one occurring in late January. The very old crescent Moon passes less than 3° north of the morning star on 25 September. The following day, Mercury and Venus are a little over 3° apart. Venus is getting quite low on the eastern horizon for all observers and will be lost to view late next month.

EARTH reaches another equinox on 23 September; spring returns to the southern hemisphere and autumn arrives in the north. The Full Moon nearest to the autumnal equinox in the northern hemisphere is traditionally called the 'Harvest Moon'; this year the Harvest Moon is about as early as it can get, occurring on 10 September. The monthly occultation of Uranus takes place on 14 September. The following day, the Moon is less than 3° south of the open cluster known the Pleiades (M45).

MARS continues to brighten as it approaches opposition later this year. Currently found in Taurus, the red planet is a brilliant magnitude −0.1 at the beginning of the month and brightens a further half a magnitude by the start of October. Although only a morning sky object for southern observers, Mars enthusiasts in northern temperate latitudes will see the red planet rise late in the evening. The phase of Mars increases from 88% to 94% illumination this month.

JUPITER returns to Pisces on the first day of the month. The Moon, just one day past full, comes to within 1.8° of the bright planet on 11 September. This is an excellent month to observe the king of the planets, with Jupiter reaching opposition on 26 September and being visible throughout the night. At magnitude −2.9, it is

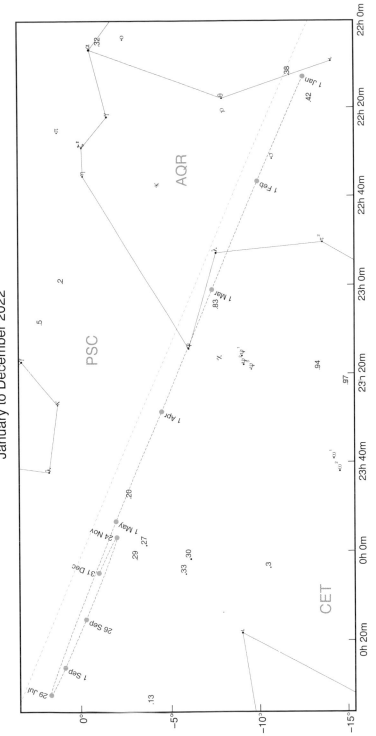

Jupiter
January to December 2022

Background stars are shown to magnitude +5.5.

Neptune in 2022

Because Jupiter and Neptune are in conjunction on 12 Apr 2022, the path of Jupiter during April is also depicted here. Neptune is indicated by blue dots and Jupiter by yellow dots.

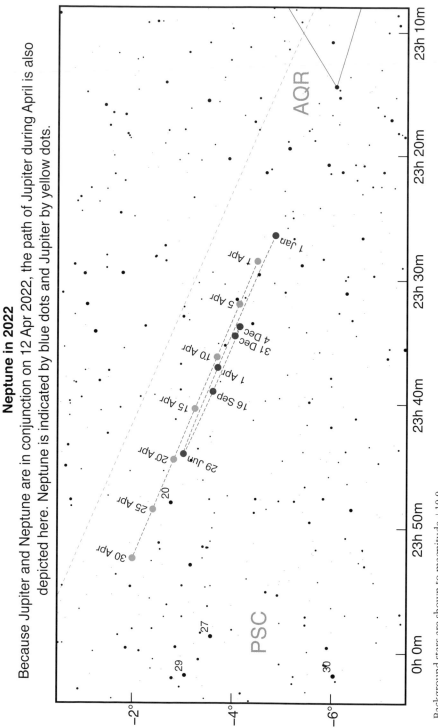

Background stars are shown to magnitude +10.0.

the brightest non-lunar body in the constellation of the Fish and a telescope reveals a large, banded 49.8 arc-second disk.

SATURN is primarily an evening sky object, having completed opposition last month. Located in Capricornus, the planet is still a bright +0.3 magnitude as Earth slowly draws away from it. It is readily observable from both hemispheres and sets in the early morning hours ahead of the Sun.

URANUS is reversing across Aries, having begun retrograde motion last month. It rises before midnight, earlier for those in northern temperate latitudes than those in the southern hemisphere, and is occulted by the waning gibbous Moon on 14 September. This event is visible across northern Africa, Europe, western and northern Asia, Iceland and Greenland beginning around 21:00 UT.

NEPTUNE reaches opposition on 16 September in Aquarius. It is at its brightest, +7.8 magnitude, at this time and appears as a tiny blue 2.5-arc-second globe in a telescope's eyepiece. It rises around sunset and sets around sunrise this month.

Early Astronomy Stamps from Brazil

Katrin Raynor-Evans

The world's first adhesive postage stamp was issued in Great Britain in 1840. Designed by Sir Rowland Hill, the Penny Black stamp produced for general public use features a profile of Queen Victoria and a star symbol in the upper left and right hand corners, whilst the lower corners are printed with letters, displaying where in the printing sheet they were positioned. Other countries soon followed suit with Brazil issuing their first country-wide stamps – *Olhos-de-Boi* or the Bulls Eye – in 1843. So, what do stamps and in particular, Brazilian stamps have to do with astronomy? Just as stamp production evolved over the years, so did the design. Stamps began to feature animals, historical figures, and transport. In 1887, Brazil introduced an astronomical element to their stamps becoming the first country to produce a stamp with an astronomy theme. Here I briefly introduce some of these early astronomy stamps issued in Brazil.

On 19 November 1889, Brazil was officially announced a republic, and a new national flag was adopted. At the time, 21 stars selected from various constellations in the Southern Celestial Hemisphere were added to the blue sphere centred on the flag, twenty stars representing each state and one representing the Neutral Municipality. Over the years, the number of states increased and the current flag, which was last modified in 1992, now depicts 27 stars – 26 representing each state and the other representing the federal district.

Featuring on the flag is the prominent asterism Crux Australis, the Southern Cross or in Portuguese, *Cruzeiro do Sul* which is located in the constellation of Crux. The five stars Acrux, Mimosa, Gacrux, Imai and Ginan form the asterism, which is in the shape of a cross,

The first astronomy stamp issued in Brazil in 1887. (Philately / Correios do Brasil)

with Acrux and Gacrux pointing the way to the South Celestial Pole. The Southern Cross is an important symbol in Brazil's political, religious, and nautical history.

The appearance of the Southern Cross on the flag is also linked to the stamps of the newly announced republic, which start to appear frequently after 1889. There is, one exception however, a stamp depicting the Southern Cross that dates to 1887 when the Empire of Brazil remained under rule by the monarchy. The blue 300 réis stamp was the first on which a pattern of stars appears and is therefore noted as the first astronomy stamp. The design of the stamp is astronomically accurate as all five stars comprising the asterism were chosen for the illustration.

Three years later in 1890, Correios (the Brazilian postal service) issued a further series of the so-called *Cruzeiro do Sul* postage stamps. Varying in denominations and colours, the image had been re-designed but still depicted the prominent five stars of the Southern Cross. In the same year, a set of newspaper stamps were issued but this time, the five-star design incorporated the famous Sugar Loaf Mountain producing a rather nice illustration of the Southern Cross shining above one of Brazil's most famous landmarks. The issue of these thematic postage stamps continued well into the 1890s proving to be a popular theme to use.

Known among philatelists as *Madrugada Republicana* – the Republican Dawn – Correios issued a series of ten stamps in 1894 to commemorate the start of the Republican period. Republican Dawn refers to a group of military officers who overthrew Emperor Dom Pedro II on 15 November 1889 with the fall of the monarchy taking place around 8am that day. Interestingly, two of the stamps issued feature a star shining brightly above the Sugar Loaf. A bit of research has shown that Venus was visible in the morning sky at the time of the coup de état affirming that the bright 'star' on the stamp is undoubtedly Venus, the morning star.

The eagle-eyed astronomer will note that on the earlier Cruzeiro do Sul stamps, the orientation of the five stars of the Southern Cross are positioned as they would be seen in the sky when looked at from Earth. However, it is normal to find in the national

A stamp depicting an architrave and the Southern Cross asterism in its reverse form was issued in 1930. (Philately / Correios do Brasil)

symbols of Brazil including the flag, that the Southern Cross is illustrated in a mirrored or reverse way as if it were being observed from space- in other words, looking down from outside the celestial sphere.

Could this mirror imagery be based on the ancient astronomical techniques and observations using an armillary sphere? After all, they were used in nautical navigation and the image of an armillary sphere featured on the Empire of Brazil flag. In 1930, the red Architrave and Southern Cross stamp issued to celebrate the fourth Pan-American Architectural Congress depicts the asterism in a mirrored form with the star Ginan now located at the left of the cross. Some collectors may think this is a mistake but in fact, it is not.

Researching early stamps issued in Brazil throughout the late nineteenth and early twentieth century provides a fascinating timeline, highlighting the country's love affair with astronomy and the influence the stars and planets have had throughout their rich history.

My thanks are due to Graham Beck, Stephen Rose and Jack Zhang for their invaluable philatelic knowledge and to Mayra Guapindaia at Correios do Brasil for supplying the images used with this article.

October

Full Moon: 9 October
New Moon: 25 October

MERCURY returns to the morning sky in what is its best dawn apparition this year for early risers in northern temperate latitudes. Mercury brightens over two magnitudes this month, from +1.4 to −1.2. It also returns to direct motion on the first of the month, passes through its ascending node the following day, and reaches perihelion for the third time in 2022 on 6 October, all before attaining greatest elongation west (only 18.0°) on the eighth. Mercury is occulted by the waning crescent Moon on 24 October beginning around 14:00 UT. This may be visible from parts of the west coast of Canada and the United States just before sunrise.

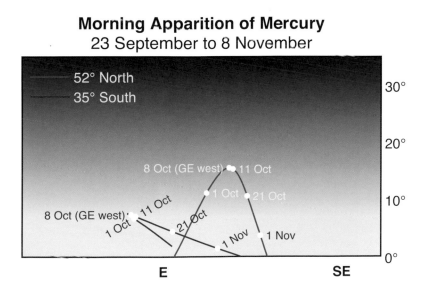

VENUS drops the last few degrees to the eastern horizon, finally undergoing superior conjunction on 22 October. Three days later it is occulted by the New Moon so it will be too near to the Sun for this event to be visible. Venus reappears as the evening star at the end of this month.

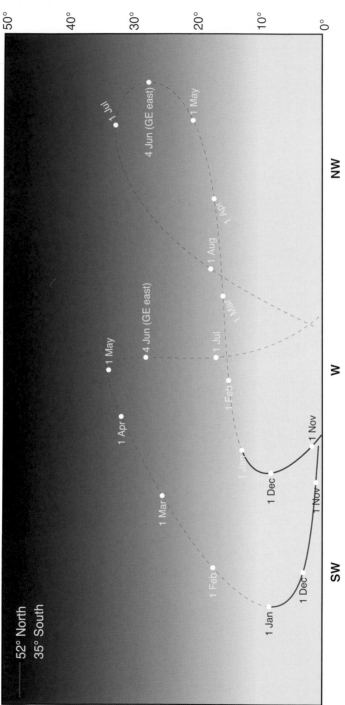

Evening Apparition of Venus
October 2022 to August 2023

52° North
35° South

EARTH enjoys three meteor showers in October, with the Draconids and Southern Taurids occurring early in the month and the Orionids in the latter half. More details are available in *Meteor Showers in 2022*. Once again the Moon occults Uranus. This happens on 11/12 October, depending on the observer's time zone. On 13 October, the Moon slides past the Pleiades (M45), appearing only 2.7° south of the open cluster. Mercury undergoes a lunar occultation on 24 October and Venus is occulted by the New Moon the next day at almost exactly the same time as a partial solar eclipse (see *Eclipses in 2022* for particulars).

MARS reaches the ascending node in its orbit about the Sun, moving from south of the ecliptic plane to north on 20 October. Ten days later it enters into retrograde motion, a sure sign that opposition is only weeks away. Located in Taurus, Mars brightens from magnitude −0.6 to −1.2 over the course of the month, outshining even nearby Aldebaran (α Tauri) and Betelgeuse (α Orionis). The red planet rises in mid-evening as seen from northern temperate latitudes and around midnight for those residing south of the equator.

JUPITER is visited by a nearly full Moon on 8 October when the two heavenly bodies are just over 2° apart in the sky. Found in the fishy constellation of Pisces, Jupiter is now past opposition and visible in the evening sky after sunset. It dims slightly as Earth leaves it behind but is still a bright magnitude −2.8 by the end of the month.

SATURN returns to direct motion across Capricornus this month. The rings, which have been opening since their minimum aspect early in June, reach a maximum tilt of 15.3° on 22 October, after which they begin to close again. This, along with Saturn receding from Earth, causes the planet to dim slightly, from magnitude +0.4 at the beginning of October to +0.6 at the end.

URANUS appears in Aries and rises not long after sunset. It is occulted by the waning gibbous Moon on 12 October, beginning around 04:30 UT. It is visible from north eastern Asia, Alaska, north western Mexico, western and north-central United States, most of Canada, Greenland and Iceland. However, the green ice giant is only sixth-magnitude so a telescope will be necessary to see it against the glare of the nearly full Moon.

NEPTUNE is now past opposition and is visible in the evening skies as soon as it gets dark. It is located in Aquarius which is well-placed for telescopic observation in both hemispheres.

An Introduction to Unusual Observatory Domes
Mills Observatory

Katrin Raynor-Evans

Astronomical observatories date back many centuries and are often considered as locations from which our ancient ancestors observed the night sky and the solstices. Examples of such observatories include Stonehenge in England and Newgrange in Ireland. These stone monuments are thought to have been used to study the sky, taking measurements and observations of the planets and their motions.

Today, modern ground-based astronomical observatories refer to small scale personal spaces in which telescopes are kept securely in observers' gardens or in larger buildings that may form part of a research facility or isolated observatories built for public use and enjoyment, which often serve to provide scientific data.

Built in 1935, Mills Observatory remains open today as a public observatory, and boasting a fully working, 25-foot diameter papier-mâché dome. (Ken Kennedy)

The dome or domes form one of the main components of the observatory. Sheltering the telescope(s) and equipment inside, domes at the larger facilities such as Palomar Observatory, California are often mechanical and can rotate 360 degrees. Known as hemispherical domes, they can easily be opened and the shutter in the dome roof positioned, exposing the sky for observations to be made.

Historically, observatory domes were mainly constructed using metals such as copper or iron. In the nineteenth century, the unusual use of papier-mâché for dome construction was introduced. This was a cheaper option compared to that of using metals but more importantly they were advantageous in limiting the weight on the supporting observatory walls. The technique of using papier-mâché was taken from that of paper-boat manufacturing – boats such as canoes and row boats were

Manufacturer of linen and twine, and amateur astronomer, John Mills (1806–1889) who left money in his will for the construction of Mills Observatory. (John McCue)

made using a wooden frame, layers of paper and adhesive. This principle was applied to designing and constructing the domes.

The first papier-mâché dome was manufactured over 140 years ago when paper-boat manufacturers E. Waters & Son of Troy, New Jersey, constructed a dome for a local observatory in 1878. Made of thick linen paper and a wooden framework, the dome could not be made as one complete piece. Instead, individual sections had to be made and bolted together. For the domes to withstand exposure to the weather, they were often completed with layers of thick cloth and lead.

Every astronomical observatory has a fascinating story to tell. I aim to present some of these observatories and their unusual domes in future articles, from the first aluminium dome built in Pittsburgh by amateur astronomer Leo Scanlon, to the Porter Turret Telescope in Springfield, Vermont where the telescope mount also serves as the dome. First, I will briefly introduce the papier-mâché dome that still exists today at Mills Observatory in Dundee, Scotland. It is only one of two papier-mâché domes surviving in the UK today, the other being at Godlee Observatory in Manchester.

Overlooking the River Tay and perched on the summit of Balgay Hill sits Mills Observatory, Britain's first public observatory. Shielded by trees, the building is made from sandstone and the seven metre diameter dome is constructed out of papier-mâché.

The observatory was built in 1935 at the bequest of John Mills, a linen and twine manufacturer and amateur scientist. It is said that Mills was greatly influenced in his life by Reverend Thomas Dick, a fellow Dundonian who was a minister and astronomer. Throughout his career, Dick believed that every city should have a public observatory. It was Mills who realised Dick's belief and left money for the building of Mills Observatory in his will.

The observatory was designed by the City Architect, James MacLellan Brown in collaboration with the Astronomer Royal for Scotland, Professor Ralph Sampson. The hand-operated rotating papier-mâché dome (originally constructed from low-density wooden fibreboard) and its steel supports were designed by Grubb Parsons and Company Optical Works, Newcastle-Upon-Tyne. The 18-inch Newtonian reflector telescope housed within the observatory at the time was also manufactured and supplied by Grubb Parsons.

After many years of use, the steel framework had corroded, and in the 1970s the dome was repaired. The original coating is still in place albeit under three layers of polymer which had been applied to waterproof the dome in 1974, 1988 and again in 2014. In 2017 the dome failed once more – its shutter and mechanism had become inoperable and could not be opened. With part of the observatory temporarily closed to the public, in 2018 a specialist conservation company was contracted, and restoration and reparations of the dome took place.

The shutter was removed and sent to Glasgow for repair whilst the remainder of the dome was repaired and restored including the winding mechanism and the dome running gear. Interestingly, a canvas/linen screen operates across the shutter opening allowing observations to be made during the daytime, keeping the glare from affecting the telescope. This was also replaced during the restoration work.

Mills Observatory re-opened in November 2018 and continues to attract thousands of visitors each year. The unique Category B listed building has a varied and interesting past shaping Scotland's long, rich history in the subject of astronomy.

Further Reading

Ford H, The Mills Observatory Dundee, *Journal of the British Astronomical Association* vol. 93, no. 6 (1983)

November

Full Moon: 8 November
New Moon: 23 November

MERCURY vanishes from the morning sky early in the month, undergoing superior conjunction on 8 November. The planet normally passes either to the north or south of the solar disk but this time, it travels directly behind the Sun. Mercury moves through its descending node the following day. Now visible in the west after sunset, Mercury is best seen from the southern hemisphere although it does not gain much altitude. The tiny planet dims slightly as the month progresses, beginning at magnitude −1.2 and ending at −0.6. The final aphelion of the year takes place on 19 November, followed two days later by a close approach to the evening star when the two objects are 1.3° apart in the sky. Another lunar occultation takes place on 24 November but this is only visible during daylight hours in Antarctica.

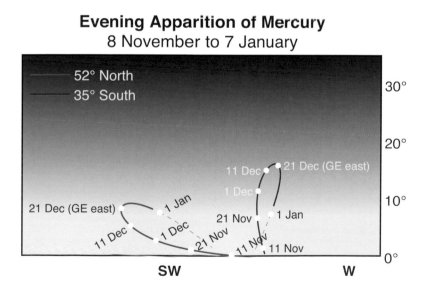

VENUS returns to the west after sunset but remains very low to the horizon as seen from northern temperate latitudes. Observers farther south have slightly better views. Passage through its descending node and a close encounter with Mercury

Uranus
January to December 2022

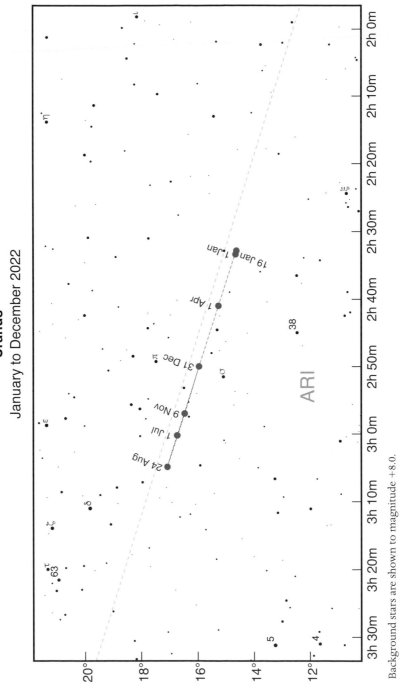

Background stars are shown to magnitude +8.0.

occur on 21 November. The day-old Moon passes 2.3° south of the evening star three days later. Venus shines at magnitude −3.9 all month and its appearance in a telescope is of a small (10 arc-seconds at most) nearly full disk.

EARTH sees a total lunar eclipse on 8 November at the same time Uranus is eclipsed by the Full Moon! For observing details of the eclipse, see *Eclipses in 2022*. The following day, the waning gibbous Moon moves 2.7° south of the open cluster M45 (the Plciades or 'Seven Sisters'). The Northern Taurid meteor shower may be affected by the light of the waning gibbous Moon but the Leonids peak around the time of New Moon (see *Meteor Showers in 2022* for more information). Finally, a lunar occultation of Mercury on 25 November takes place during daylight hours.

MARS is looping back through Taurus this month, brightening from magnitude −1.2 to −1.8 and growing in size from 15.1 arc-seconds to 17.2 arc-seconds in telescopes. The waning gibbous Moon passes 2.5° north of the red planet on 11 November. Mars is best seen in the winter skies of the northern hemisphere where it rises in the early evening.

JUPITER is visible after the sun has vanished, not setting until well after midnight. Blazing away at magnitude −2.7 in faint Pisces, it is well-placed for observations from both hemispheres. The waxing gibbous Moon appears 2.4° south of the giant planet on 4 November and twenty days later, Jupiter ceases retrograde motion.

SATURN reaches east quadrature on 11 November. Shining at magnitude +0.6 in Capricornus, Saturn is an evening sky object. It sets before midnight for those in north temperate latitudes but southern observers retain views of the ringed planet until midnight or a little afterwards.

URANUS is occulted by an eclipsed Full Moon on 8 November. This unusual event begins around 12:30 UT and is visible from most of Asia, Alaska, and western and northern Canada (although not Vancouver Island). The following day Uranus reaches opposition in the constellation of Aries. It is shining at its brightest, +5.6 magnitude, and appears as a 3.7-arc-second green disk in the eyepiece of a telescope. The most distant of the naked-eye planets rises at sunset and disappears with the sunrise.

NEPTUNE is visible in the evening and early morning skies in Aquarius this month. At eighth-magnitude, a small telescope will be necessary to see the far-away planet.

The Biggest Meteorite in the World

Susan Stubbs

Imagine an object that weighs approximately six times as much as a bull African elephant and you have the mass of the Hoba West Meteorite. The largest known intact single piece of meteorite on Earth, the Hoba West Meteorite was found on a farm of the same name in the Otjozondjupa region of the Otavi Mountains of Namibia in 1920 by the farmer Jacobus Hermanus Brits. Located around 20 kilometres west of the city of Grootfontein, it has never been moved from where it fell because of its enormous mass, estimated to be in the region of 60 tonnes. Hoba means 'gift' in the local Khoekhoegowwab dialect, and this object has proved to be so in that it is a significant tourist attraction for the local area.

The Hoba impact left no crater, and is thought to have impacted Earth about 80,000 years ago. It is unusual as it has two flat surfaces and it is thought that it

The Hoba West Meteorite as a tourist attraction following its excavation. Thousands visit it each year to marvel at its size. (commons.wikimedia.org/wiki/File:Hoba_meteorite,_Grootfontein3. jpg/Creative Commons License 2.0/Flickr/Damien du Toit)

did not break up before impact as it was slowed by the Earth's atmosphere to such an extent that it hit the ground at terminal velocity of around 0.32 kilometres per second, having slowed down from 10 kilometres per second when it entered the atmosphere. As there is no crater it was discovered by chance by the owner of the land whilst ploughing. It is the most massive naturally-occurring piece of iron on Earth, so must have caused major damage to his ploughing implements! It measures 2.7 metres $\times$ 2.7 metres $\times$ 0.9 metres and is embedded in Kalahari Limestone, overlaying granite. Analysis of its 30 centimetre oxide-shale crust suggests that the local climate in the past was wetter, with an annual rainfall of between 500 and 750 mm, in contrast to 450 mm today. The meteorite fell during the Pleistocene Era, at a time when much of the UK was covered by ice sheets and the first humans were beginning to evolve in Africa and spread across the world. Occupying an area of 6.5 square metres, the Hoba West Meteorite is nearly twice the weight of its nearest rivals El Chaco (37 tonnes) and Ahnighito, the largest fragment of the Cape York meteorite (31 tonnes).

Meteorites are pieces of rock from objects such as comets, asteroids and meteoroids, or even the Moon or Mars, which enter the Earth's atmosphere and reach the surface of our planet. Historically, meteorites have been classified in three ways: stony (94%), iron (5%) and stony-iron (1%). Stony meteorites are thought to be leftover fragments from the formation of the Solar System, and include those from the Moon and Mars. Iron meteorites are comprised mostly of iron and nickel and are likely to come from asteroid cores that melted early in their history. The Hoba is in this category, comprising 84% iron and 16% nickel along with traces of other metals, mainly cobalt. Stony-iron meteorites contain both rock and metals and originate from asteroid collisions and mixing of material from the cores and surfaces. Around 500 tonnes of meteoritic debris falls to Earth every day and out of this around 10 meteorites of reasonable size will actually be found and identified. The impact of a large meteorite may result in the formation of a crater, and there are around 350 proven (or possible) impact structures on Earth, the largest of which is the 1.2 kilometre diameter Meteor Crater in Arizona.

It is surprising that there is no crater, as something so massive would be expected to come through our atmosphere at high speed and impact with enormous energy. It is thought that it skipped along the surface, and hence didn't bury itself in a crater because it fell at a lower speed due to the drag of its flat shape. There is an

The Hoba West Meteorite, photographed during the first Geological expedition in 1932. Dr. L. J. Spencer is second from the left. (Wikimedia Commons/Flickr/American Museum of Natural History Library/Guide Leaflet 1901)

abundance of iron oxides in the surrounding soil, and this suggests it was once much larger than 60 tonnes. Erosion, sampling and vandalism have decreased its mass from a probable original 66 tonnes to the current 60.

The first scientific report on the Hoba West Meteorite was compiled in 1932 by geologist Leonard James Spencer (1870–1959), and chemist and mineralogist Max Hutchinson Hey (1904–1984), of the Mineral Department of The British Museum (Natural History) in London. The first published photograph was taken by Friedrich William Kegel. Recovery to extract the high nickel content was considered, but fortunately abandoned, although it has been much studied by scientists in the intervening years, sadly contributing to the loss of the original mass! There are a number of pieces to be found around the world, with the largest, at 4.2 kilograms, in Philadelphia, USA. The presence of radioactive nickel has allowed estimation of age, which is thought to be between 190 million and 410 million years.

The Hoba is now a tourist attraction. In 1955 it was declared a national monument by Namibia to protect it from vandalism, and in 1979 this was extended to cover the land on which it sits. In 1987 Mr. J. Engelbrecht, the then owner of the land, donated the meteorite and the surrounding area to the state, and a visitor centre opened shortly afterwards. Thousands now visit annually to see this marvel from space, and to be awed by its size. Jacobus Hermanus Brits would be astounded by the popularity of his find!

December

Full Moon: 8 December
New Moon: 23 December

MERCURY ends the year in the west, starting at a bright magnitude −0.6 and ending at +1.3. It reaches its maximum declination south for 2022 on 8 December. Greatest elongation east (20.1°) occurs on 21 December, followed by passage through its ascending node eight days later. It is just 1.4° north of Venus at the same time. Mercury is best observed from the southern hemisphere; it is brightest at the beginning of the month but highest in the sky mid-month. Inferior conjunction awaits early next year.

VENUS reaches its maximum declination south for the year on 13 December and encounters Mercury for the third time in 2022 on 29 December. The evening star shines at magnitude −3.9 and is best observed from equatorial latitudes. However, it is gaining useful height every day and observers in both hemispheres will have a decent evening apparition next year.

EARTH reaches solstice on 21 December. Summer officially begins in the southern hemisphere and winter grips the north. The final 2022 lunar occultation of Uranus takes place on 5 December, followed three days later by an occultation of the red planet. Between these two events, on 6 December, the Moon passes 2.7° south of the star cluster known as the Pleiades (M45). Two major meteor showers grace terrestrial skies this month, the Geminids mid-month and the Ursids about a week later. More details are available in the section *Meteor Showers in 2022*.

MARS is at its minimum distance from Earth, 0.5447 au, on the first day of the month. Five days later it reaches its maximum declination north for the year. Opposition occurs on 8 December when Mars glows a brilliant magnitude −1.9 and appears as a reddish disk of diameter 17.3 arc-seconds in the telescope. The Full Moon gets in on the act, occulting Mars on the same day. The event begins around 03:00 UT which is actually the previous evening in the parts of Mexico, the United States and Canada where the occultation is visible. The event may also be seen in Greenland, Iceland and much of western Europe, including Britain and Ireland, before sunrise on 8 December.

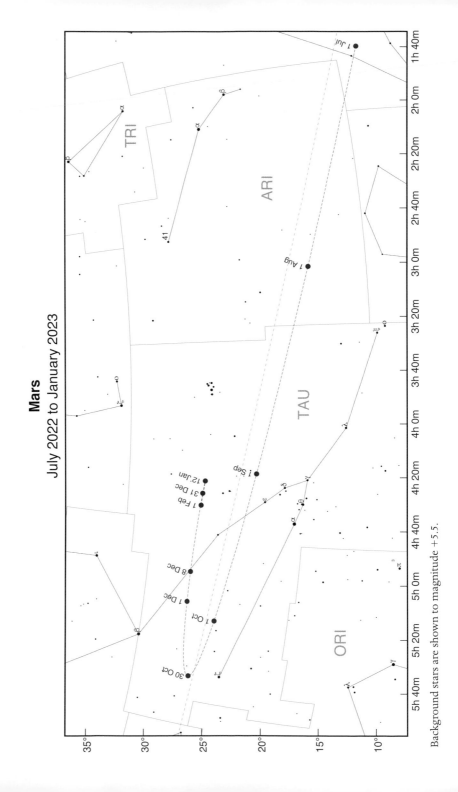

Mars

July 2022 to January 2023

Background stars are shown to magnitude +5.5.

JUPITER is visited by the Moon twice this month, on both the second day of December and on 29 December. In both instances the Moon is a little over 2° south of Jupiter. The largest planet in the solar system reaches east quadrature on 22 December which presents astrophotographers opportunities to capture some interesting shadow interplay between the planet and its Galilean satellites. Found in Pisces, Jupiter is an evening sky object, not setting until around midnight or a little after.

SATURN is visible in the evening, setting mid-evening for those in north temperate latitudes and late evening for those south of the equator. Located the Capricornus, Saturn slowly dims a tenth of a magnitude over the course of the month, ending at +0.8.

URANUS is well-placed for observation in the evening hours. Located in Aries, it does not set until well after midnight. It undergoes one final lunar occultation this year when, on 5 December, it vanishes behind the disk of the waxing gibbous Moon. This event is visible from just after sunset in northern Africa, northern parts of the Middle East, all of Europe and northern Asia. Because of the brightness of the nearly full Moon, a telescope will be necessary to spot the sixth-magnitude object.

NEPTUNE returns to direct motion on 4 December and reaches east quadrature ten days later. Currently found in Aquarius, it is already aloft by the time night falls and sets around midnight. The waxing crescent Moon approaches within 3° of Neptune on 28 December; next year the Moon will make increasingly close passes by Neptune, eventually occulting the faint planet.

The Next 'Small Step'

David M. Harland

The first phase of human exploration of the Moon was wrapped up fifty years ago with the final Apollo lunar mission in December 1972.

-oOo-

For thousands of years, people gazed up at the Moon and wondered what it was. Some even imagined visiting it. With the dawning of the Space Age, and the competitive spirit of the Cold War, conducting such a voyage became a real prospect.

In 1961 President John F. Kennedy challenged his nation to achieving the goal, before the decade was out, of landing a man on the Moon. In July 1969 Neil A. Armstrong achieved this "giant leap for Mankind" by taking a "small step" off the foot pad of the Apollo 11 lander *Eagle* onto the dusty regolith of the Sea of Tranquillity.

In December 1972, the sixth Apollo lunar landing drew the program to a close with Eugene A. Cernan and Harrison H. Schmitt, who was known as Jack, setting *Challenger* down in a valley nestled in the Taurus Mountains.

On their final day, they drove their Lunar Roving Vehicle up one of the mountains to an elevation of 75 metres. Upon looking back across the valley, Cernan was awed by their achievement. "You know, Jack, when we finish, we'll have covered this valley from corner to corner!"

"That was the idea," Schmitt replied pointedly.

Having driven for a total of 35 kilometres and collected 110 kg of samples, it was time to pack up and leave the Moon. Prior to following Schmitt up into *Challenger*, Cernan, who would later claim the moniker of the Last Man on the Moon, paused for a moment to reflect on the significance of what he was about to do.

"As I take Man's last step from the surface, I would like to just say what I believe history will record – that America's challenge of today has forged Man's destiny of tomorrow. And as we leave the Moon … we leave as we came and, God willing, as we shall return, with peace and hope for all Mankind." With that, he headed up the ladder for the final time.

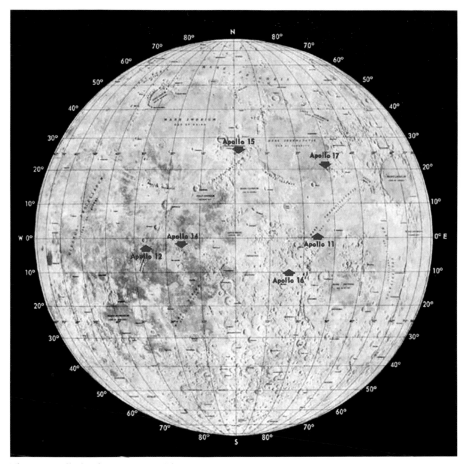

The six Apollo landing sites. (NASA)

Shortly after Cernan and Schmitt had rejoined their colleague Ron Evans in their mother craft, *America*, President Richard Nixon issued a statement that was read up to the astronauts. "As *Challenger* leaves the surface of the Moon, we are conscious not of what we leave behind, but of what lies before us." After saying that Mankind makes progress by pursuing its dreams, he pointed out, "This may be the last time in this century that men will walk on the Moon." Although he insisted "space exploration will continue," he seemed oblivious to the fact that he had dashed the dreams of the next generation of astronauts, eager to renew and expand lunar exploration.

The ascent stage of *Challenger* lifts off. (NASA)

On 31 December 1999, National Public Radio interviewed Arthur C. Clarke, who was famous for his prowess in forecasting so many of the twentieth century's most fundamental developments. Asked whether there was anything particular in the preceding 100 years that he never could have anticipated, Clarke replied, "Yes, absolutely. The one thing I never would have expected is that, after centuries of wonder and imagination and aspiration, we would have gone to the Moon – and then stopped!"

The Moon, as one school child put it to me around that time, "was a place no one goes any more".

But by the turn of the twenty-first century, NASA was flying robotic probes to survey the Moon from orbit with sensors far more sophisticated than those that were available to Apollo. One discovery was that at the lunar poles there are elevated terrains that are almost continually in sunlight and craters whose permanently shadowed floors are laced with water ice. The mantra of in-situ resource utilisation changed the prospects for renewing lunar exploration. A polar base would not only have almost uninterruptible solar power but also water for drinking and conversion into oxygen and hydrogen rocket propellants. In addition, with suitable apparatus it would be feasible to extract metals from the minerals in the regolith and use the

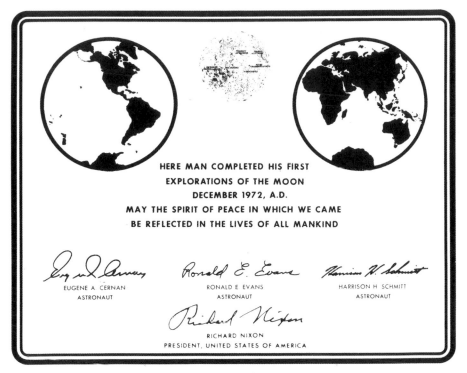

HERE MAN COMPLETED HIS FIRST
EXPLORATIONS OF THE MOON
DECEMBER 1972, A.D.
MAY THE SPIRIT OF PEACE IN WHICH WE CAME
BE REFLECTED IN THE LIVES OF ALL MANKIND

EUGENE A. CERNAN
ASTRONAUT

RONALD E. EVANS
ASTRONAUT

HARRISON H. SCHMITT
ASTRONAUT

RICHARD NIXON
PRESIDENT, UNITED STATES OF AMERICA

The commemorative plaque on the descent stage of *Challenger*. (NASA)

waste as a construction material. And this is not just wishful thinking. An optimistic NASA is now working to return people to the Moon in the next several years with the Artemis Program.[1] The plan is for the first crew to include a woman, who will, almost certainly, become the next person to take that 'small step' onto the lunar surface.

1. In mythology, Artemis was the sister of Apollo.

Comets in 2022

Neil Norman

Comets are generally regarded as being the most unpredictable objects of the Solar System and may appear at any given moment with no prior warning. The next 'great comet' could in fact be discovered at any time, and it is advisable to keep a close eye on social media or the British Astronomical Association (BAA) Comet Section page **www.ast.cam.ac.uk/~jds** where the latest comet discoveries are announced.

Best Prospects For 2022

At the time of writing, a total of 76 comets are due to return to perihelion during 2022, with 71 of these belonging to the Jupiter family of comets, four being long-period and one a lost comet (these are always worth trying to find in case of a outburst or an asteroid being found in the predicted position). Although only three (see below) are expected to attain magnitude 10 or brighter, comets are famed for their unpredictability, and outbursts could occur leading to one or more of them becoming brighter at any time.

19P/Borrelly

Comet 19P/Borrelly was discovered by French astronomer Alphonse Louis Nicolas Borrelly on 28 December 1904, and is a Jupiter family comet with an orbital period of 6.8 years. On 21 September 2001 this object was visited by Deep Space 1, which passed the nucleus at a distance of around 2,200 kilometres. Images taken by the probe revealed the nucleus to be shaped rather like a bowling pin with a mean diameter of just 2.4 kilometres.

DATE	RA	DECLINATION	MAGNITUDE	CONSTELLATION
1 Jan 2022	00 18 43	−18 11 38	9.2	Cetus
15 Jan 2022	00 46 49	−08 06 26	9.1	Cetus
1 Feb 2022	01 26 17	+04 24 15	9.1	Pisces
15 Feb 2022	02 01 52	+14 08 00	9.2	Aries
1 Mar 2022	02 41 17	+22 52 54	9.5	Aries
15 Mar 2022	03 25 06	+30 17 33	9.8	Aries
1 Apr 2022	04 24 17	+37 08 36	10.3	Perseus
15 Apr 2022	05 17 00	+40 52 23	10.7	Auriga

19P/Borrelly imaged on 5 October 2015 by Denis Buczynski (BAA) from Tarbatness Observatory, Ross-shire, Scotland. (Denis Buczynski)

4P/Faye

A brief mention should be made of this comet as it was also visible when the *Yearbook of Astronomy* first came out in 1962 (see below). It was discovered by French astronomer Hervé Auguste Étienne Albans Faye on 23 November 1843. Another Jupiter family comet, during its last apparition in 2014, it underwent an outburst of two magnitudes.

DATE	RA	DECLINATION	MAGNITUDE	CONSTELLATION
1 Jan 2022	06 36 28	+07 15 48	10.9	Monoceros
15 Jan 2022	06 26 55	+08 15 48	11.2	Monoceros
1 Feb 2022	06 22 02	+09 55 58	11.7	Monoceros
15 Feb 2022	06 24 30	+11 21 13	12.1	Monoceros
1 Mar 2022	06 32 12	+12 36 46	12.6	Monoceros

4P/Faye imaged on 26 February 2015 by Denis Buczynski (BAA) from Tarbatness Observatory, Ross-shire, Scotland. (Denis Buczynski)

104P/Kowal

Discovered on 13 Jan 1979 by American astronomer Charles Thomas Kowal, this comet was actually first seen in 1973 by Leo Boethin. Another Jupiter family comet, 104P/Kowal has a diameter of 2 kilometres and orbital period of 5.90 years which takes it from an aphelion distance of 5.35 au to perihelion on 11 January 2022 when it will approach to within 1.18 au of the Sun.

DATE	RA	DECLINATION	MAGNITUDE	CONSTELLATION
1 Jan 2022	00 07 59	−05 26 07	9.2	Pisces
15 Jan 2022	01 07 10	−01 46 19	9.2	Cetus
1 Feb 2022	02 07 10	+03 51 41	9.2	Cetus
15 Feb 2022	03 47 15	+08 31 32	9.5	Taurus
1 Mar 2022	05 00 31	+12 06 29	10.0	Orion
15 Mar 2022	06 05 53	+14 09 06	10.6	Orion

C/2017 K2 (PanSTARRS)

This comet was discovered on 21 May 2017 when at a distance of over 16 au from the Sun, with precovery images taking it back as far as 2013 when its solar distance was 23 au. Although a very large object, with a nucleus in the order of at least 10 kilometres in diameter, it will not approach the Sun any closer than 1.8 au in late December this year, and is on a hyperbolic trajectory that will eventually see it kicked out of the Solar system.

DATE	RA	DECLINATION	MAGNITUDE	CONSTELLATION
1 Jan 2022	17 58 46	+13 15 12	10	Ophiuchus
15 Jan 2022	18 10 06	+12 21 43	10	Ophiuchus
1 Feb 2022	18 23 27	+11 39 19	10	Ophiuchus
15 Feb 2022	18 33 37	+11 21 13	10	Ophiuchus
1 Mar 2022	18 42 37	+11 15 54	9	Ophiuchus
15 Mar 2022	18 49 55	+11 20 30	9	Aquila
1 Apr 2022	18 55 38	+11 33 15	8	Aquila
15 Apr 2022	18 56 50	+11 42 30	8	Aquila
1 May 2022	18 53 04	+11 39 02	7	Aquila
15 May 2022	18 44 02	+11 08 07	7	Ophiuchus
1 Jun 2022	18 24 32	+09 26 18	6	Ophiuchus
15 Jun 2022	18 01 20	+06 43 16	6	Ophiuchus
1 Jul 2022	17 29 12	+01 53 19	6	Ophiuchus
15 Jul 2022	17 00 17	−03 34 29	6	Ophiuchus

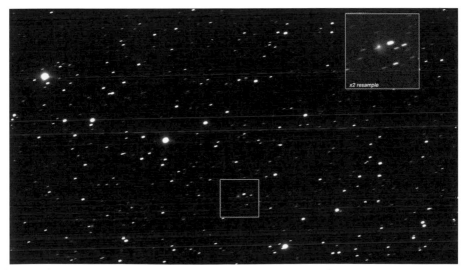

This image of C/2017 K2 (PanSTARRS) was captured by Nick James (BAA) from Chelmsford, Essex, UK on 25 May 2020 and reveals the comet as a 15th magnitude object located around 9 au from the Sun. (Nick James)

A Look Back at 1962

As you are aware, the *Yearbook of Astronomy* is celebrating its 60th birthday, the first edition having appeared in 1962, so I thought it would be interesting to take a look back at some of the comets the readers of the first edition had to look forward to in that year.

The first comet was 18D/Perrine-Mrkos which came to perihelion in February 1962, and was originally discovered by the American-Argentine astronomer Charles Dillon Perrine on 9 December 1896. For some time it was thought to be a fragment of Biela's Comet (for more about Biela's Comet see the article *Biela's Comet: A Tale of Two Parts* elsewhere in this volume) and was considered lost after the 1909 appearance, but was rediscovered by the Czech astronomer Antonín Mrkos on 19 October 1955. As far as the 1962 apparition was concerned, it only went on to reach magnitude 17.5 and was last seen at the following apparition of 1968 and since then has been lost.

Comet 41P/Tuttle-Giacobini-Kresak reached perihelion in March and attained magnitude 10 with a trace of tail visible. Discovered by American astronomer Horace Parnell Tuttle on 3 May 1858, and re-discovered independently by French astronomer Michel Giacobini and Slovak astronomer Ľubor Kresák in 1907 and 1951 respectively, 41P/Tuttle-Giacobini-Kresak is a member of the Jupiter family of comets, approaching to within 0.27 au of Earth on 12 April 1962.

Comet 41P/Tuttle-Giacobini-Kresak imaged near Messier 108 and Messier 97 (Owl Nebula) on 22 March 2017. (Wikimedia Commons/Kees Scherer)

Discovered in 1843 by Hervé Faye from Paris Observatory, during a close passage by the Earth around a month after it had passed perihelion, 4P/Faye attained magnitude 13 during its 1962 apparition, during which it was observed to have a short tail. Its current (2022) apparition follows perihelion passage on 8 September 2021.

Comet C/1962 C1 (Seki-Lines) was discovered by Richard D. Lines and his wife Helen on 4 February, a few hours before an independent discovery made by Japanese astronomer Tsutomu Seki. This comet went on to become a naked-eye comet in March, and eventually reached its highest magnitude of −1 on 27 March. During April, brightness dropped to magnitude 3 but the

Hervé Faye (Wikimedia Commons)

Comet C/1962 C1 Seki-Lines, photographed on the evening of 9 April 1962 from Kōchi, Japan by co-discoverer Tsutomu Seki. (Tsutomu Seki)

tail grew to 20 degrees. By month's end the comet had dropped to magnitude 7 with only a trace of tail remaining.

The final comet of 1962 was C/1961 R1 (Humason), discovered by American astronomer Milton L. Humason on 1 September 1961. This comet is remembered for its later showing in 1962 when it appeared blue in colour images. The comet was a large one with an apparent magnitude of 1.5-3.5 and with an estimated nucleus size of between 30 and 40 kilometres. The comet moved through perihelion on 10 December 1962 at a solar distance of 2.13 au, well beyond the orbit of Mars. It was brightest in November 1962 at magnitude 5.5 while still some 2.17 au from the Sun.

Comet C/1961 R1 (Humason) as imaged from Palomar Observatory, California on 1 September 1962. Located in Capricornus at time of the image, some 2.13 au from the Sun, the comet showed a very dynamic ion tail, the likes of which had not been seen since the appearance of Comet Morehouse in 1908. (Wikipedia/C.E. Kearns/K. Rudnicki)

Minor Planets in 2022

Neil Norman

Minor planets are a collection of varying sized pieces of rock left over from the formation of the Solar system around 4.6 billion years ago. Millions of them exist, and to date almost 800,000 have been seen and documented, with around 470,000 having received permanent designations after being observed on two or more occasions. Different family types of asteroids exist also, Amor asteroids which are defined by having an orbital period of over 1 year and also an orbit that does not cross that of the Earth. Apollo asteroids have their perihelion distances within that of the Earth and thus can approach us to within a close distance and Trojan asteroids have their home at Lagrange points both 60 degrees ahead and behind the planet Jupiter respectively. These asteroids pose no problems to the Earth.

Most of these objects reside within the asteroid belt that lies between the orbits of Mars and Jupiter. However, some asteroids have orbits which allow them to interact with major planets, including the Earth. Over 20,000 asteroids have orbits that can bring them into close proximity to our planet. Such an object is referred to as a Near Earth Asteroid (NEA). To be classed as an NEA, an object must have an orbit which allows it to pass within a distance of 1.3 au of Earth. A Potentially

This image of the Near Earth Asteroid 1998 OR2 taken on 20 April 2020 shows motion over a period of 1 hour. This object passed within 3.9 million miles (6.3 million kilometres) of our planet on 29 April 2020. (Northolt Observatory, London, England)

Hazardous Asteroid (PHA) on the other hand is defined as being one which can approach to within 0.05 au (19.5 lunar distances) of our planet, and have a diameter of at least 140 metres (460 feet).

The following is a list of objects that passed close to Earth during 2019 and 2020.

DATE OF CLOSEST APPROACH	DISCOVERED	OBJECT	LUNAR DISTANCE	DISCOVERER
8 Jan 2019	8 Jan 2019	2019 AS5 *	0.039	Mt. Lemmon
17 Jan 2019	16 Jan 2019	P10LGkb **	0.088	Pan-STARRS
1 Mar 2019	1 Mar 2019	2019 EH1 **	0.061	CATALINA
4 Mar 2019	5 Mar 2019	CO9Q 4H2 *	0.069	Mt. Lemmon
5 Sep 2019	6 Sep 2019	2019 RP1 *	0.10	CATALINA
31 Oct 2019	31 Oct 2019	2019 UN13 **	0.015	CATALINA
16 Aug 2020	16 Aug 2020	2020 QG *	0.024	Zwicky Transient Facility
18 Aug 2020	18 Aug 2020	2020 QF2 **	0.545	Mt. Lemmon
20 Aug 2020	20 Aug 2020	2020 QY2 **	0.173	Mt. Lemmon
2 Sep 2020	2 Mar 2011	2011 ES4 ***	0.65	Mt. Lemmon

*   Discovered after closest approach.
**   Discovered less than 24 hours before closest approach.
*** Discovered more than a year before closest approach.

Lunar Distance is a reference used to determine the distance more easily of an approaching asteroid to Earth. The distance to the Moon is (on average) 384,400 km (238,855 miles), as an example, if an asteroid was to approach the Earth to a lunar distance of 0.50, this would equal 192,200 kilometres (119,427 miles).

The light trail of asteroid 2020 QG captured by the Zwicky Transient Facility at Palomar Observatory, California on 16 August 2020. The image was taken six hours after closest approach as the object was heading away from Earth. (ZTF/Caltech Optical Observatories)

The following list details a selection of the brightest minor planets visible in binoculars or small telescopes during 2022.

1 Ceres

This object was discovered by Italian astronomer Giuseppe Piazzi from Palermo Observatory, Sicily on 1 January 1801. Ceres is the largest object in the asteroid belt with a diameter of 945 kilometres (587 miles). Initially believed to be a planet, in the 1850s Ceres was reclassified as an asteroid following the discoveries of many other objects in similar orbits. The gap in the table (from April to November) is due to Ceres being in solar conjunction and (as seen from Earth) too close to the Sun to view safely.

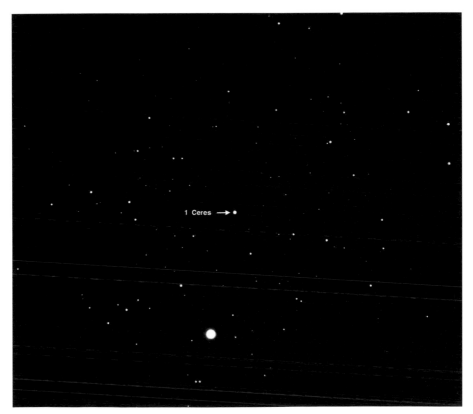

This image of Ceres was captured on 24 July 2019. The bright star near the 6 o'clock position is fifth-magnitude Lambda (λ) Librae. (Loren C. Ball, Emerald Lane Observatory (843), Decatur, Alabama, USA)

DATE	R.A.	DEC	MAG	CONSTELLATION
1 Jan 2022	03 46 49	+17 43 32	7.8	Taurus
15 Jan 2022	03 43 11	+18 27 09	8.0	Taurus
1 Feb 2022	03 46 10	+19 35 28	8.3	Taurus
15 Feb 2022	03 54 10	+20 41 02	8.5	Taurus
1 Mar 2022	04 06 23	+21 50 44	8.7	Taurus
15 Mar 2022	04 22 07	+23 00 28	8.8	Taurus
1 Apr 2022	04 45 03	+24 19 20	8.9	Taurus
15 Nov 2022	11 31 54	+12 41 01	8.7	Leo
1 Dec 2022	11 54 09	+11 17 17	8.6	Leo
15 Dec 2022	12 11 30	+10 24 07	8.5	Virgo

3 Juno

With a diameter of around 230 kilometres (143 miles), Juno is the eleventh largest asteroid and one of the two largest stony (S-type) asteroids. Discovered by German astronomer, Karl Ludwig Harding on 1 September 1804, its mean distance from the Sun is 2.6 au, its journey around our star taking 4.36 years to complete.

DATE	R.A.	DEC	MAG	CONSTELLATION
15 Aug 2022	23 16 16	+00 04 49	8.8	Pisces
1 Sep 2022	23 06 40	−02 37 50	8.3	Pisces
15 Sep 2022	22 56 33	−05 22 31	8.2	Aquarius
1 Oct 2022	22 46 28	−08 25 05	8.5	Aquarius
15 Oct 2022	22 42 01	−10 29 15	8.8	Aquarius
1 Nov 2022	22 44 10	−11 55 23	9.1	Aquarius
15 Nov 2022	22 52 14	−12 11 56	9.3	Aquarius
1 Dec 2022	23 07 26	−11 37 55	9.5	Aquarius
15 Dec 2022	23 24 57	−10 29 25	9.7	Aquarius

4 Vesta

Discovered by German astronomer Heinrich Wilhelm Matthias Olbers on 29 March 1807, Vesta is one of the largest of the asteroids, with a diameter of 525 kilometres (326 miles) and an orbital period of 3.63 years. Vesta holds the distinction of being the brightest minor planet visible from Earth.

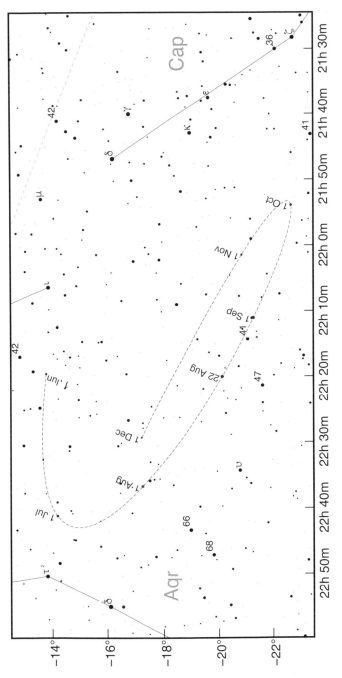

Vesta in 2022

This chart depicts the path of minor planet 4 Vesta. The red dots show the position of Vesta on the first day of each month and at opposition on 22 August. They are scaled to indicate the magnitude, which varies from 7.1 on 1 June to 5.7 on 22 August to 7.8 on 1 December. Stars are shown to magnitude 9.0, which is the limit for binoculars or a small telescope. (David Harper)

DATE	R.A.	DEC	MAG	CONSTELLATION
1 Jun 2022	22 19 09	−13 46 59	7.1	Aquarius
15 Jun 2022	22 32 03	−13 36 23	7.0	Aquarius
1 Jul 2022	22 40 08	−14 05 20	6.7	Aquarius
15 Jul 2022	22 42 04	−15 10 02	6.5	Aquarius
1 Aug 2022	22 36 04	−17 15 11	6.1	Aquarius
15 Aug 2022	22 26 01	−19 14 20	5.8	Aquarius
1 Sep 2022	22 10 07	−21 20 35	5.9	Aquarius
15 Sep 2022	21 59 08	−22 21 18	6.3	Aquarius
1 Oct 2022	21 53 02	−22 37 12	6.7	Capricornus
15 Oct 2022	21 53 07	−22 07 58	7.0	Capricornus
1 Nov 2022	22 01 05	−20 48 01	7.4	Aquarius
15 Nov 2022	22 12 08	−19 17 15	7.6	Aquarius
1 Dec 2022	22 29 07	−17 10 32	7.8	Aquarius
15 Dec 2022	22 47 00	−15 03 27	8.0	Aquarius

5 Astraea

Discovered by the German amateur astronomer Karl Ludwig Hencke on 8 December 1845, Astraea has a diameter of 119 kilometres (74 miles) and orbits the Sun once every 4.12 years. Another S-type object, Astraea has a highly reflective surface, which suggests that it is composed of a mixture of nickel-iron with silicates of magnesium and iron.

DATE	R.A.	DEC	MAG	CONSTELLATION
1 Aug 2022	23 36 43	−04 53 30	11.8	Aquarius
15 Aug 2022	23 30 54	−05 58 43	11.5	Aquarius
1 Sep 2022	23 19 16	−07 44 13	11.1	Aquarius
15 Sep 2022	23 07 51	−09 16 32	11.1	Aquarius
1 Oct 2022	22 55 46	−10 46 10	11.4	Aquarius
15 Oct 2022	22 48 20	−11 37 08	11.7	Aquarius
1 Nov 2022	22 44 58	−11 57 48	12.0	Aquarius
15 Nov 2022	22 47 08	−11 41 59	12.2	Aquarius

6 Hebe

The second of two asteroids discovered by German amateur astronomer Karl Ludwig Hencke, Hebe first came to light on 1 July 1847, and is most likely to be the parent body of the H-type ordinary chondrites, the most common type of meteorite and which account for about 40% of all meteorites hitting Earth.

Karl Ludwig Hencke as depicted in the January 1891 issue of *Sidereal Messenger* (Wikimedia Commons / Sidereal Messenger)

DATE	R.A.	DEC	MAG	CONSTELLATION
1 Oct 2022	07 58 43	+10 43 19	10.5	Cancer
15 Oct 2022	08 18 32	+09 44 05	10.5	Cancer
1 Nov 2022	08 38 16	+08 38 55	10.3	Cancer
15 Nov 2022	08 50 10	+08 00 31	10.2	Cancer
1 Dec 2022	08 57 53	+07 46 10	10.0	Cancer
15 Dec 2022	08 58 39	+08 10 23	9.7	Cancer

7 Iris

The first of ten asteroids discovered by the English astronomer John Russell Hind, Iris was first spotted on 13 August 1847. Iris is of a stony composition and ranks as the fourth brightest object in the asteroid belt.

DATE	R.A.	DEC	MAG	CONSTELLATION
1 Jan 2022	07 53 50	+15 36 13	7.9	Gemini
15 Jan 2022	07 38 16	+15 29 13	7.7	Gemini
1 Feb 2022	07 21 30	+15 36 52	8.3	Gemini
15 Feb 2022	07 13 47	+15 48 29	8.7	Gemini
1 Mar 2022	07 12 58	+15 58 46	9.1	Gemini
15 Mar 2022	07 18 32	+16 03 05	9.5	Gemini
1 Apr 2022	07 32 07	+15 54 42	9.9	Gemini
15 Apr 2022	07 47 28	+15 33 21	10.2	Gemini
1 May 2022	08 08 10	+14 50 48	10.4	Gemini

8 Flora

Another of the asteroids discovered by John Russell Hind, Flora has a mean diameter of 128 kilometres (80 miles) and first came to light on 18 October 1847. The name of this asteroid was suggested by John Frederick William Herschel in honour of the Roman goddess of flowers and gardens.

DATE	R.A.	DEC	MAG	CONSTELLATION
1 Jan 2022	13 30 16	−03 35 00	11.5	Virgo
15 Jan 2022	13 45 06	−04 24 34	11.4	Virgo
1 Feb 2022	13 58 41	−04 49 51	11.1	Virgo
15 Feb 2022	14 05 10	−04 38 36	10.9	Virgo
1 Mar 2022	14 06 30	−03 57 20	10.6	Virgo
15 Mar 2022	14 02 10	−02 48 26	10.3	Virgo
1 Apr 2022	13 49 58	−01 01 14	10.0	Virgo
15 Apr 2022	13 36 24	+00 26 21	9.8	Virgo

27 Euterpe

Euterpe was discovered by John Russell Hind on 8 November 1853 and is one of the brightest of the asteroids. Named after the Muse of music and lyric poetry in Greek mythology, Euterpe has an orbital period of 3.59 years and a diameter of around 124 kilometres (77 miles). This object is the parent body of the stony inner-belt Euterpe family of asteroids, which has a total of around 400 members.

DATE	R.A.	DEC	MAG	CONSTELLATION
15 Aug 2022	03 00 50	+15 12 50	11.2	Aries
1 Sep 2022	03 19 44	+16 19 17	10.9	Aries
15 Sep 2022	03 30 56	+16 52 31	10.6	Taurus
1 Oct 2022	03 37 08	+17 05 43	10.2	Taurus
15 Oct 2022	03 35 32	+16 55 17	9.8	Taurus
1 Nov 2022	03 24 45	+16 17 14	9.2	Taurus
15 Nov 2022	03 11 12	+15 33 35	8.8	Aries
1 Dec 2022	02 56 17	+14 50 06	9.3	Aries
15 Dec 2022	02 48 26	+14 36 21	9.7	Aries

Meteor Showers in 2022

Neil Norman

Everybody loves to see a shooting star dash across the sky and on any given night you can expect to see several of these. Quite often these will be 'sporadic' meteors, that is to say they can appear from any direction and at any time during the observing session. These meteors arise when a meteoroid – perhaps a particle from an asteroid or a piece of cometary debris orbiting the Sun – enters the Earth's atmosphere and burns up harmlessly high above our heads, leaving behind the streak of light we often refer to as a "shooting star". The meteoroids in question are usually nothing more than pieces of space debris that the Earth encounters as it travels along its orbit, and range in size from a few millimetres to a couple of centimetres in size. Meteoroids that are large enough to at least partially survive the passage through the atmosphere, and reach the Earth's surface without disintegrating, are known as meteorites.

At certain times of the year the Earth encounters more organised streams of debris that produce meteors over a regular time span and which seem to emerge from the same point in the sky. These are known as meteor showers. These streams of debris follow the orbital paths of comets, and are the scattered remnants of comets that have made repeated passes through the inner solar system. The ascending and descending nodes of their orbits lie at or near the plane of the Earth's orbit around the Sun, the result of which is that at certain times of the year the Earth encounters and passes through a number of these swarms of particles.

The term 'shower' must not be taken too literally. Generally speaking, even the strongest annual showers will only produce one or two meteors a minute at best, this depending on what time of the evening or morning that you are observing. One must also take into account the lunar phase at the time, which may significantly influence the number meteors that you see. For example, a full moon will probably wash out all but the brightest meteors.

The following is a table of the principle meteor showers of 2022 and includes the name of the shower; the period over which the shower is active; the Zenith Hourly Rate (ZHR); the parent object from which the meteors originate; the date of peak shower activity; and the constellation in which the radiant of the shower is located. Most of the information given is self-explanatory, but the Zenith Hourly Rate may need some elaborating.

The Zenith Hourly Rate is the number of meteors you may expect to see if the radiant (the point in the sky from where the meteors appear to emerge) is at the zenith (or overhead point) and if observing conditions were perfect and included dark, clear and moonless skies with no form of light pollution whatsoever. However, the ZHR should not be taken as gospel, and you should not expect to actually observe the quantities stated, although 'outbursts' can occur with significant activity being seen.

The observer can make notes on the various colours of the meteors seen. This will give you an indication of their composition; for example, red is nitrogen/oxygen, yellow is iron, orange is sodium, purple is calcium and turquoise is magnesium. Also, to avoid confusion with sporadic meteors which are not related to the shower, trace the path back of the meteor and if it aligns with the radiant you can be sure you have seen a genuine member of the particular shower.

Meteor Showers in 2022

SHOWER	DATE	ZHR	PARENT	PEAK	CONSTELLATION
Quadrantids	1 Jan to 5 Jan	120	2003 EH$_1$ (asteroid)	3/4 Jan	Boötes
Lyrids	16 Apr to 25 Apr	18	C/1861 G1 Thatcher	22/23 Apr	Lyra
Eta Aquariids	19 Apr to 28 May	30	1P/Halley	6/7 May	Aquarius
Tau Herculids	19 May to 19 Jun	Varies	73P/Schwassmann–Wachmann	9 Jun	Hercules
Delta Aquariids	12 Jul to 23 Aug	20	96P/Machholz	28/29 Jul	Aquarius
Perseids	17 Jul to 24 Aug	80	109P/Swift-Tuttle	12/13 Aug	Perseus
Draconids	6 Oct to 10 Oct	10	21P Giacobini-Zinner	8/9 Oct	Draco
Southern Taurids	10 Sep to 20 Nov	5	2P/Encke	10 Oct	Taurus
Orionids	2 Oct to 7 Nov	20	1P/Halley	21/22 Oct	Orion
Northern Taurids	20 Oct to 10 Dec	5	2004 TG$_{10}$ (asteroid)	12 Nov	Taurus
Leonids	6 Nov to 30 Nov	Varies	55P/Tempel/Tuttle	17/18 Nov	Leo
Geminids	4 Dec to 17 Dec	75+	3200 Phaethon (asteroid)	13/14 Dec	Gemini
Ursids	17 Dec to 26 Dec	10	8P/Tuttle	22/23 Dec	Ursa Minor

Quadrantids

The parent object of the Quadrantids has been identified as the near-Earth object of the Amor group of asteroids 2003 EH1 which is likely to be an extinct comet. The radiant lies a little to the east of the star Alkaid in Ursa Major and the meteors are fast moving, reaching speeds of 40 km/s. Maximum activity occurs over a short period of just 2–3 hours, a thin crescent Moon having set in the early evening making way for dark skies and what should be a really good show.

This image of a Quantrandid meteor was taken in January 2020 from Maryland, USA. (Mike Hankey)

Lyrids

Produced by particles emanating from the long-period comet C/1861 G1 Thatcher, these are very fast moving meteors that approach speeds of up to 50km/s. The peak falls on the night of 22/23 April with the radiant lying near the prominent star Vega in the constellation of Lyra. This year some of the fainter meteors may be drowned out by the light from a last quarter Moon.

Eta Aquariids

One of the two showers associated with 1P/Halley, the Eta Aquariids are active for a full month between 19 April and 21 May. The radiant lies just to the east of the star Sadalmelik (α Aquarii), from where up to 30 meteors per hour are normally expected during the period of peak activity which occurs in the pre-dawn skies of 7 May. A waxing crescent moon will have set early in the previous evening, leaving dark skies for what should be a decent show.

Tau Herculids

Appearing to originate from the star Tau (τ) Herculis, this shower runs from 19 May to 19 June with peak activity taking place on 9 June. The Tau Herculids were first recorded in May 1930 by observers at the Kwasan Observatory in Kyoto, Japan. The parent body has been identified as the comet 73P/Schwassmann–Wachmann, discovered on 2 May 1930 by the German astronomers Arnold Schwassmann and

Comet 73P/Schwassmann-Wachmann imaged from Namibia, South Africa on 26 March 2017 during its perihelion passage of that year. This rapidly fragmenting comet is responsible for the Tau Herculids stream. (Gerald Rhemann)

Arno Arthur Wachmann during a photographic search for minor planets being carried out from Hamburg Observatory in Germany.

73P/Schwassmann–Wachmann has an orbital period of 5.36 years. During 1995 the comet began to fragment, and by the time of its 2006 return, at least eight individual fragments were observed (although the Hubble Space Telescope spotted dozens more). 73P/Schwassmann–Wachmann appears to be close to total disintegration. The observed rate of meteors from this shower are low, but with this said, recent research suggests that Earth will be getting closer to the streams of debris left behind by the fragmenting comet and thus heavier meteor showers can be expected in the coming years with another hefty shower possible for 2049. Peak activity in 2022 occurring shortly after the first quarter moon should ensure a reasonably good display, although it may be worthwhile looking for enhanced activity up to two weeks before the peak date. This will be during the period when 73P/Schwassmann–Wachmann is approaching its next perihelion passage – due on 25 August 2022 – and an earlier outburst may take place.

Delta Aquariids

Linked to the short-period sungrazing comet 96P/Machholz, the Delta Aquariids coincide with the more prominent Perseids, although Delta Aquariid meteors are generally much dimmer than those associated with the Perseids, making identification somewhat easier. The radiant lies to the south of the Square of Pegasus, close to the star Skat (δ Aquarii). Located to the north of the bright star Fomalhaut in Pisces Austrinus, the radiant is particularly well placed for those observers situated in the southern hemisphere. The shower peaks during the early hours of 29 July, at the time of new moon, leaving dark skies and the potential for a rewarding display.

Perseids

Associated with the parent comet 109P/Swift-Tuttle, the Perseids are a beautiful sight, with meteors appearing as soon as night falls and up to 80 meteors per hour often recorded. This is usually one of the best meteor showers to observe, large numbers of very bright meteors often being seen, although in 2022, at the time of peak activity on the night/morning of 12/13 August, the moon will be just past full phase and will block out all but the brightest meteors.

Perseid meteor captured near Ursa Major on 11 August 2018 from Oxfordshire, England. (Mary McIntyre)

Southern Taurids/Northern Taurids

Running collectively from 10 September to 10 December, the Taurids are two separate showers, with southern and northern components, the Southern Taurids linked to the periodic comet 2P/Encke and the Northern Taurids to the asteroid 2004 TG10. The southern hemisphere encounters the first part of the stream, followed later by the northern hemisphere encountering the second part. The ZHR of these showers is low (between 5 and 10 per hour), although they can be beautiful to watch as they glide across the sky. As the Southern Taurids peak on 10 October the moon will be at full phase drowning out the fainter meteors. The Northern Taurids fare little better, the waning gibbous moon on 12 November also hiding many of the fainter shower members.

Draconids

Also known as the Giacobinids, the Draconid meteor shower emanates from the periodic comet 21P/Giacobini-Zinner. The duration of the shower is just four days from 6 October to 10 October, with the shower peaking on 8/9 October. The ZHR of this shower varies with poor displays in 1915 and 1926 but stronger displays in 1933, 1946, 1998, 2005 and 2012. Radiating from the "head" of Draco, the meteors from this shower travel at a relatively modest 20 km/s. The Draconids are generally quite faint, and a full moon this year will block out all but the brightest meteors and make for difficult observing.

Orionids

The second of the meteor showers associated with 1P/Halley, the Orionids radiate from a point a little to the north of the star Betelgeuse in Orion. Best viewed in the early hours when the constellation is well placed, the shower takes place between 2 October and 7 November with a peak on the night of 21/22 October. The velocity of the meteors entering the atmosphere is a speedy 67 km/s. This year, a thin and waning crescent moon will leave dark skies and a potentially rewarding display.

Leonids

This is a fast moving shower with particles varying greatly in size and which can create lovely bright meteors that occasionally attain magnitude −1.5 (or about as bright as Sirius) or better. The radiant is located a few degrees to the north of the bright star Regulus in Leo. The parent of the Leonid shower is the periodic comet 55P/Tempel-Tuttle which orbits the Sun every 33 years. It was last at perihelion in 1998 and is due to return in May 2031. The Zenith Hourly Rates vary due to the Earth encountering material from different perihelion passages of the parent comet. For example, the storm of 1833 was due to the 1800 passage, the 1733 passage was responsible for the 1866 storm and the 1966 storm resulted from the 1899 passage (for additional information see Courtney Seligman's article *Cometary Comedy and Chaos* in the *Yearbook of Astronomy 2020*).

The shower peaks on the night of 17/18 November, and a moon just past last quarter will hide many of the fainter Leonid meteors.

Geminids

The Geminid meteor shower was originally recorded in 1862 and originates from the debris of the asteroid 3200 Phaethon. Discovered in October 1983, this rocky five kilometre wide Apollo asteroid made a relative close approach to Earth on 10 December 2017, when it came within 0.069 au (10.3 million kilometres/6.4 million miles) of our planet. The Geminid radiant lies near the bright star Castor in Gemini and the shower peaks on the night of 13/14 December. This

A montage of 44 meteors, 35 of which are Geminids, imaged on 13 December 2008 from Castellammare di Stabia, Italy. (Ernesto Guido)

is considered by many to be the best shower of the year, and it is interesting to note that the number of observed meteors appears to be increasing annually. However, this year a waning gibbous moon will block many of the fainter meteors.

Ursids

Discovered by William F. Denning during the early twentieth century, this shower is associated with comet 8P/Tuttle and has a radiant located near Beta (β) Ursae Minoris (Kochab). With relatively low speeds of 33 km/s, the Ursids are seen to move gracefully across the sky. The shower runs from 17 December to 26 December with peak activity taking place on the night of 22/23 December when the new moon will leave dark skies and potential for a good show.

Ursid bolide captured on 22 December 2008 from Castellammare di Stabia, Italy. (Ernesto Guido)

Article Section

Astronomy in 2021

Rod Hine

"There's Gold in Them Thar Hills!"

This was a phrase in American popular culture, variously misquoted, attributed to a Mark Twain character and even put into song by Frankie Marvin in 1931. But where did the famous gold come from to inspire countless prospectors to join in the "Gold Rush"? It has long been understood that elements with atomic numbers greater than iron had been forged in the heart of supernova explosions since this is the only way to get the extreme energies necessary for atomic fusion to produce such elements. However, detailed consideration of the mechanism by which gold could be formed by neutron capture during supernova nucleosynthesis appears to show that there is far too much gold in "them thar hills" to have been so produced. Stars massive enough to fuse gold during their supernova event generally become black holes after their explosion and so most of the gold gets sucked into the black hole, according to Chiaki Kobayashi of the University of Hertfordshire.

Kobayashi is the lead author in a paper published in *The Astrophysical Journal* in September 2020 which examined in impressive detail the evolution of all the elements from carbon to uranium from first principles.[1] By pulling together data from no fewer than 341 other papers and using the latest theoretical models for nucleosynthesis, she and co-authors Amanda Karakas of Monash University, Australia, and Maria Lugaro of the Hungarian Academy of Sciences, Budapest, were able to account for the origins of many elements with satisfying accuracy. For instance, they showed how neutron star collisions would produce strontium, a result that was confirmed by the August 2017 observation by LIGO of the gravitational waves from a neutron star merger. Combined with simultaneous observations from VIRGO and an intense gamma-ray burst observed by NASA's Fermi Gamma-ray Space Telescope, the event named GW170817 was eventually traced to a source in NGC4993 some 130 million light years from Earth.

Optical observations in the infra-red of the new light source, led by Marcelle Soares-Santos of Brandeis University, Massachusetts, revealed that huge amounts of heavier metals had been produced and thrown into space. Co-author Edo

---

1. For further details see **iopscience.iop.org/article/10.3847/1538-4357/abae65**

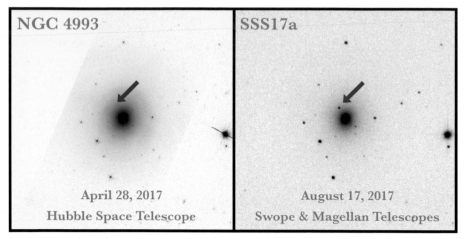

An image (right) taken on 17 August 2017 with the Swope Telescope at the Las Campanas Observatory in Chile shows the light source generated by a neutron-star merger in the galaxy NGC 4993. In the image (left) taken on 28 April 2017 with the Hubble Space Telescope, the neutron star merger has not occurred and the light source, known as SSS17a, is not visible. (arxiv. org/pdf/1710.05452.pdf/D.A. Coulter, et al.)

Berger of Harvard University said, "There's about 10 times the Earth's mass in gold and platinum alone". So, given the fact that neutron star mergers are shown to be very efficient at synthesizing the heavier elements such as gold, more so than conventional supernovae, can this account for the high levels of gold in the Earth's core and crust?

Not so, according to Kobayashi. Her detailed examination of the sources of the elements showed that there are just not enough neutron star mergers to account for it all. In any case, most of the gold in the Earth is locked away in the core along with iron and other metals from the local supernova that must have given birth to the Earth and the Solar System. The gold sought by the Gold Rush prospectors most likely was added to the crust and mantle by planetesimals and by asteroid impacts in the Late Heavy Bombardment period, some four billion years ago.

As of now, the origin of gold remains something of an enigma. Either there are some fundamental flaws in the modelling or there is some other unknown process that spews gold into space. There is still plenty of work for astrophysicists to do before the source of all that wonderful glittery, precious metal is explained.[2]

2. To read the article 'Identification of strontium in the merger of two neutron stars' published in *Nature*, 23 October 2019, visit **www.nature.com/articles/s41586-019-1676-3**

Research Sparked by a Tweet

Time was when researchers gathered around the blackboard or the water cooler to discuss wild ideas, but it is perhaps a sign of the times that a recent paper published in *The Astrophysical Journal* was sparked by a casual comment in a Tweet. The background to this paper was reported in the popular science blog "Bad Astronomy" on the Syfy.com website. According to blog author Phil Plait, it started with a Tweet from Judy Schmidt, which included an image of IC5063, in which she speculated about the faint appearance of "cones" on either side of the nucleus. IC5063 is an active galaxy about 160 million light years distant and the images in question were taken by the venerable Hubble Space Telescope in November 2019 during a survey of nearby active galactic nucleus (AGN) host galaxies in red and near infra-red wavelengths. That survey was led by Aaron J. Barth of the University of California, Irvine, USA. Schmidt herself normally works in image processing on Hubble and Chandra data at University of Maryland, USA.

A side-by-side view of a cleaned version of the galaxy IC 5063 (left) with one processed to bring out the detail of the dark rays (right). (NASA, ESA, STScI, Judy Schmidt)

In an increasingly curious series of Tweets, speculation about the faint ray-like shadows eventually reached astronomer Peter Maksym of the Harvard-Smithsonian Centre for Astrophysics, who studies the behaviour and evolution of black holes at the centres of galaxies. A team led by Maksym worked with Schmidt to process and analyse the images and did indeed reveal the

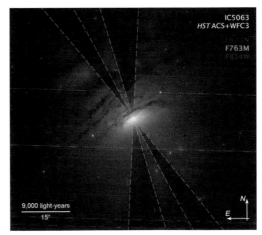

An annotated version of the IC 5063 image showing the dark rays more clearly. The cameras and filters used (one red (673 nanometers) and two in the near infrared (763 and 814 nm) are listed in the upper right. (NASA, ESA, STScI and W. P . Maksym (CfA))

dark conical rays, analogous to the "crepuscular rays" often seen in the light of the setting sun. Comparing Schmidt's initial "discovery image" with the final processed images shows the amazing power of digital processing in the hands of experts.

In the final paper, published in September 2020, Maksym et al explain in detail just what can be learned from the images to understand the complex environment close to the nucleus of IC5063, with perhaps a warped dusty toroidal disk structure that could cause the shadowing effects.[3]

The Vera C. Rubin Observatory

Originally known as the Large Synoptic Survey Telescope Observatory (LSST) and now renamed to honour the famous woman astronomer Vera Rubin, this telescope will be seeing "first light" in 2021 and promises to change dramatically the kind of observations available to astronomers. Its unique combination of a huge 8.4 metre mirror, three-mirror anastigmat optical design and 3.2 Gigapixel prime focus camera, will enable it to photograph the whole sky every few nights. Located on a 2,682 metre high mountain in northern Chile alongside the existing Gemini South and Southern Astrophysical Research Telescopes it will cover most of the southern hemisphere sky at wavelengths from 330 nm (near ultraviolet) to 1080 nm (near infra-red). The field of view is an amazing 3.5 degrees in diameter – so the sun or moon images will fit across the field seven times over![4]

Central to its success will be the CCD imaging instrument installed at prime focus, the largest digital camera ever built. The sensor is a 64 centimetre diameter mosaic of 189 individual 16 megapixel detectors which will be cooled to 173 K ($-100\,^{\circ}$C) in a vacuum chamber to reduce noise levels. Each pixel is a mere 0.2 arc-seconds wide. Capable of taking a 15 second exposure every 20 seconds and thus generating 64 Gigabytes of raw data per image, this camera will produce far more data per night than can ever be analysed by humans. In fact, the technical problems of storing and processing all the data will likely turn out to be the most difficult and challenging part of the whole project.

In order to allow the scientific community to access this avalanche of data, it is intended to process the data according to three timescales. So-called "prompt" data will be processed within 60 seconds and alerts will be issued for anything that has changed brightness or position relative to archived images. This facility will enable mapping of small objects in the solar system, as well as giving rapid news

---

3. For further details, see **arxiv.org/abs/2009.10153**
4. More images in website **gallery.lsst.org/bp/#/folder/2334406/51622237** show the unusual design of the telescope with the multiple mirror optical system.

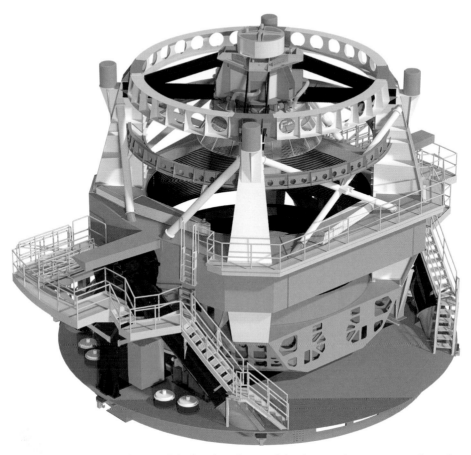

A three dimensional rendering of the baseline design of the dome with a cutaway to show the telescope. (Rubin Observatory/NSF/AURA)

of transient events and rapid follow-up of gravitational wave events and gamma-ray bursts. Further processing will be done on a daily basis and a complete annual catalogue will be released with data on billions of galaxies and stars.

Overall, the Vera C. Rubin Observatory is seen as an important resource and is well-funded by the USA National Science Foundation, the United States Department of Energy and other government bodies, along with substantial donations from billionaires Bill Gates and Charles Simonyi. It remains to be seen whether the increasing light pollution caused by satellite constellations such as Starlink will prove detrimental to the long-term aims of the project.

The Australian Square Kilometre Array Pathfinder Radio Telescope (ASKAP)

Just as the Vera C. Rubin Observatory promises rapid all-sky surveys in the optical wavelengths, ASKAP is already delivering incredibly rapid surveys of the radio sky and has even discovered a totally new class of radio objects. Operated by the Commonwealth Scientific and Industrial Research Organisation (CSIRO), ASKAP is a network of 36 antennas, each 12 metres in diameter, located in a remote part of the Western Australia outback that has been under construction for some years. The unique feature of ASKAP is the use of "Phased Array Feeds" which give each dish the capability to form 36 beams rather than the single beam feed typical of conventional radio telescopes.

When all the dishes became fully operational, the array was used to produce a survey of the whole sky visible from the location at Murchison, Western Australia. In just 300 hours of observations, around 83% of the observable universe was mapped in great detail, comprising 903 individual images each with 70 billion pixels. "For the first time, ASKAP has flexed its full muscles, building a map of the universe in greater detail than ever before, and at record speed," lead study author

The ASKAP array consists of 36 dish antennas, each 12 metres in diameter, and spread out to give baselines of up to 6 kilometres. The array is located in a designated radio-quiet zone of Western Australia. (Wikimedia Commons/CSIRO)

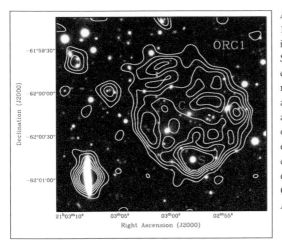

ASKAP radio contour map of ORC 1 superimposed on a visible and near infrared image from the Dark Energy Survey (DES). The object C at the centre may be a galaxy but has no radio emission, whereas the object S is a bright radio source and is probably a star-forming galaxy. Further observations would be needed to establish if these are related to the diffuse ORC object. (Ray P. Norris et al "Unexpected Circular Radio Objects at High Galactic Latitudes" Arxiv:2006.14805v1)

David McConnell, a CSIRO astronomer, said in a statement. "We expect to find tens of millions of new galaxies in future surveys."

All the data will be available publicly through CSIRO's Data Access Portal and will no doubt provide a treasure trove for researchers around the world.

ASKAP has already revealed an entirely new class of radio objects that have been named "ORCs", or "Odd Radio Circles". They are described by CSIRO astronomer Ray Norris and his colleagues in a paper submitted to *Nature Astronomy*.[5] The new objects are approximately circular disks which do not appear to correspond to known objects or even possible imaging artefacts. Three of these new objects, named ORC1, 2 and 3, were found in the ASKAP data and a further similar one, ORC4, was discovered in the archive of images of the galaxy cluster Abell 2142 taken by the Giant Metrewave Radio Telescope (GMRT) near Pune, India.

After describing the objects, the authors discuss in great detail many other possible causes ranging from artefacts due to processing and supernova remnants to galactic planetary nebulae and cluster halo effects. They conclude that ORCs are something new and worth investigating.

Arecibo Collapse

As a sad counterpoint to the new instruments that have come on-stream recently, or that are planned for completion soon, the catastrophic collapse of the 305 metre (1000 feet) diameter Arecibo Radio Telescope in December 2020 has deprived

5. See also the article 'Unexpected Circular Radio Objects at High Galactic Latitude' in *Publications of the Astronomical Society of Australia* at **arxiv.org/pdf/2006.14805.pdf**

Weakened by failures of suspension cables in August and November, the receiver assembly (pictured here in 2019) finally collapsed into the dish on 1 December 2020. (Wikimedia Commons / Mariordo / creativecommons.org / licenses / by-sa / 4.0 / deed.en)

astronomers of a famous and valuable resource. Completed in 1963, it was the largest radio telescope in the world for more than five decades and researchers using it have made countless contributions in many different fields of astronomy, from pulsars and comets to SETI. It was also used in radar astronomy experiments and even used in 1974 to transmit the controversial "Arecibo Message", inspired by astronomer and astrophysicist Frank Drake[6], to the globular star cluster M13 in November 1974.

---

6. To learn more about the work of Frank Drake, see the article 'Frank Drake and His Equation' by David M. Harland elsewhere in this volume.

Solar System Exploration in 2021

Peter Rea

This article was written in the autumn/winter of 2020. As all the missions mentioned are active, the status of some missions may change after the print deadline.

Moon Missions in Context

The Apollo moon landing missions came to an end in December 1972. There have been very few landing missions to the moon since then, although orbiting missions have continued at a slow pace. The decade of the 1970s saw two Russian missions to land successfully on the lunar surface. Then nothing until 2008 and 2009 with two impactors followed by two landers in 2013, Chang'e 3 and 2019 Chang'e 4 both from the Chinese. At the time of writing Chang'e 5 is due for launch in late 2020 for a soft landing. This lack of moon landing missions is due to change in the next few years as the Americans build up to a crewed return to the Moon in around 2024 with their Artemis program. Before humans once again leave their footprints on the Moon a series of robotic landers are set to launch towards Earth's nearest neighbour.

Return to the Moon's Surface

The Commercial Lunar Payload Services is a NASA program to contract out to private industry a means of delivering relatively small science payloads to the lunar surface particularly near the lunar south pole. A main mission objective is to look for natural lunar resources. It further seeks to test In Situ Resource Utilization (ISRU) concepts which are to process materials found on the moon that would usually be brought from Earth. The first four missions selected are the Peregrine Lander from Astrobotic Technology and the Nova-C lander from Intuitive Machines, and currently scheduled for launch in 2021. These two will be followed by the XL-1 lander from Masten Space Systems in 2022 and the Volatiles Investigating Polar Exploration Rover (VIPER) delivered to the Moon's surface by the Griffin lander from Masten Space Systems in 2023. As this is a rapidly developing program, readers are advised to obtain the latest information by searching "Commercial Lunar Payload Services" or visit their website shown below. Further links to some of the early missions can be found below.

Commercial landers will carry NASA provided science and technology payloads to the lunar surface, paving the way for NASA astronauts to land on the Moon by 2024. (NASA)

CLPS website: **www.nasa.gov/content/commercial-lunar-payload-services**
NASA's Moon to Mars Program: **www.nasa.gov/moontomars**
Astrobotic Website: **www.astrobotic.com**
Intuitive Machines: **www.intuitivemachines.com**
Masten Space Systems: **www.masten.aero**

Hope for the UAE at Mars

At 22.58 GMT the 19 July 2020 a Japanese H2A rocket launched the United Arab Emirates first interplanetary mission on a 7-month journey to Mars. Called HOPE (Arabic *Al Amal*) it was built under the direction of the Mohammed Bin Rashid Space

Centre in Dubai with technical assistance from the University of Colorado in Boulder, the University of California in Berkeley, and Arizona State University. HOPE is scheduled to arrive at Mars in February 2021 where it will be placed into a highly elliptical orbit with a closest point to Mars of 20,000 kilometres and furthest point from Mars of 43,000 kilometres and a period of 55 hours. From this orbit it will study the daily and seasonal weather patterns. HOPE carries three main instruments, a camera and two spectrometers, one in the infra-red and one in the ultra-violet.

An artist's impression of the United Arab Emirates HOPE Mars mission spacecraft in flight configuration. (Mohammed Bin Rashid Space Centre)

The Mohammed Bin Rashid Space Centre website with details of the Hope mission can be found here: **www.mbrsc.ae/emirates-mars-mission**

Mission website can be found here: **www.emiratesmarsmission.ae**

China's First Mars Mission

On 23 March 2020, a Long March 5 rocket lifted off from the Wenchang launch site to place the Tianwen-1 mission on an interplanetary trajectory toward Mars. Tianwen-1 is a combined orbiter, lander and rover. The mission plan calls for the spacecraft to enter orbit around Mars sometime in February 2021. After surveying the potential landing sites in the Utopia Planitia region of Mars the lander with rover attached will separate from the orbiter and enter the Martian atmosphere. The lander will use a combination of heat shield, parachute, retrorockets and airbags for the soft landing reminiscent of the American Mars Exploration Rovers. The landing attempt is expected around two months after orbital insertion.

Still Persevering with Mars Rovers

Lifting off from launch complex 41 at Cape Canaveral on 30 July 2020 an Atlas 5 with tail number AV-088 sent the Mars 2020 rover named Perseverance on a seven month journey to Mars. The 1050kg, $2.7 billion, six-wheeled rover will arrive at Mars on 18 February 2021. Its main goal is to search for past life on Mars in the 49 kilometre diameter Jezero crater located on the boundary between the Syrtis Major and Isidis Planitia. Jezero crater is the site on an ancient river delta and it is toward this scientifically significant feature that the Perseverance rover will be directed. It carries a suite of instruments to study the local geology and a total of 23 cameras to study the surface features and record the entry decent and landing

NASA's Mars 2020 rover will store rock and soil samples in sealed tubes on the planet's surface for future missions to retrieve. (NASA/JPL-Caltech)

in far more detail than any previous Mars lander. An experimental helicopter called Ingenuity is carried underneath the rover for release sometime after landing. Ingenuity is solely a demonstration of technology; it is not designed to support the Mars 2020 mission. A robotic arm on the rover carries a drill to take rock samples that will be placed in special containers to be deposited on the surface of Mars to await the arrival of a future rover which will collect these sample tubes for return to Earth. NASA in partnership with the European Space Agency is working to develop between them this "fetch" rover and sample return vehicle.

The Perseverance mission website can be found at **mars.nasa.gov/mars2020**
Details on the Mars Helicopter can be found at **mars.nasa.gov/technology/helicopter**

Challenges for the Insight Mission on Mars

At the time of writing the NASA Insight mission is still having difficulties with the "Mole" that should be boring into the sub surface of Mars to gather heat flow data from the interior. The Mole should have bored its way a few meters below the surface but cannot seem to find friction with the soil to do this. As of early autumn 2020 the Mole is just below the surface, having been pushed down by the robotic arm, but little downward progress is being made. The seismic experiment which was placed onto the surface before the Mole is returning good science and has shown evidence for Marsquakes. As the recovery operations for the Mole are ongoing, readers should visit the mission website shown below for the latest information.

NASA Mars Exploration Website can be found at **mars.nasa.gov**
NASA Insight Mission can be found at **mars.nasa.gov/insight**

It's Gravity Assists All the Way to Mercury

The BepiColombo mission was launched successfully on 20 October 2018 this joint European Space Agency / Japan Aerospace Exploration Agency mission to Mercury is well on its way to the inner solar system. Placing spacecraft in orbit around Mercury can be very costly in terms of fuel requirements. However, by using the gravity of other planets to slow the spacecraft down greatly reduces the onboard fuel requirements. This process is discussed in the article "Gravity Assists – Something for Nothing?" elsewhere in this volume. The first of nine gravity assists to reduce heliocentric velocity occurred successfully on 10 April 2020. The Earth is travelling around the Sun at just over 107,000 kph and needs to shed a large portion of this to "fall" toward Mercury. The launch vehicle cannot deliver all this velocity change. The gravity assists need to deliver the rest. Readers can follow the mission and keep up to date with the other gravity assists by going to the mission website which can be found here **sci.esa.int/bepicolombo**

Artist's impression of the BepiColombo spacecraft in cruise configuration, flying past Earth on 10 April 2020. (ESA/ATG medialab)

Europe's Mission to the Sun

The Solar Orbiter mission from the European Space Agency was launched successfully on 10 February 2020 on a potential 7-year investigation of the Sun. Solar Orbiter will take just under two years to reach its initial operational orbit which will be highly elliptical with a perihelion or closest point to the Sun of 0.28 au putting it inside the orbit of Mercury. The aphelion or furthest point from the Sun of 0.91 au almost out to the orbit of the Earth. Using the gravity of Venus and

Artist's impression of ESA's Solar Orbiter spacecraft which will observe our Sun from within the orbit of Mercury at its closest approach. (ESA/ATG medialab)

the Earth the orbits inclination to the plane of the ecliptic will be raised to around 24 degrees and potentially up to 33 degrees in an extended mission enabling images to be captured of the Sun's polar region, an area that has never been imaged before.

Solar Orbiter will use a combination of 10 instruments to observe the turbulent solar surface, the Sun's hot outer atmosphere and changes in the solar wind. Specialist imaging systems will obtain high-resolution images of the Sun's atmosphere, the corona, as well as the solar disc. Other instruments will measure the solar wind and the solar magnetic field in the vicinity of the orbiter.

Mission website can be found at **sci.esa.int/solarorbiter**

Where Would We Be Without a TAGSAM?

The American OSIRIS-REx mission to the 500 metre diameter asteroid 101955 Bennu reached a pivotal point on 20 October 2020 when the long awaited attempt to gather samples for return to Earth was attempted. At just over 18 light minutes away it was not possible for engineers to intervene. The sample acquisition had to be fully automatic. The spacecraft slowly descended at a snail pace toward the asteroid

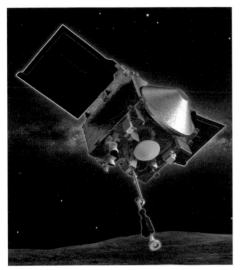

Artist's impression of NASA's OSIRIS-REx spacecraft collecting a sample from asteroid Bennu using the Touch-and-Go Sample Acquisition Mechanism (TAGSAM). (University of Arizona/NASA Goddard Space Flight Center)

This image shows the sample site Nightingale, which is located high in asteroid Bennu's northern hemisphere and visible here in the upper left of the image. The image was taken by the PolyCam camera on NASA's OSIRIS-REx spacecraft on 22 January 2020 from a distance of 600 metres. The field of view is 9.2 metres across. (NASA/Goddard/University of Arizona)

surface. A landing site had been chosen many months before and in August of 2020 a rehearsal of the procedures was carried out but stopped short of touching the asteroid. This rehearsal proved that OSIRIS-REx could perform the necessary functions out of contact with Earth. The site chosen was in the northern part of Bennu where the touchdown point was surrounded by large boulders which had to be avoided. The instrument that would pick up the samples was called the TAGSAM or Touch-And-Go Sample Acquisition Mechanism. The Americans love their acronyms! OSIRIS-REx threaded its way past the large boulders with great precision and at just after 11.00pm UTC on 20 October the TAGSAM touched Bennu and collected significant amounts of surface material. It was a mission goal to collect a minimum of 60 grams but images of the head of the TAGSAM showed far more than that. In fact it was overflowing. A flap that would snap shut to seal in the samples was prevented from closing because of so much material and images from an onboard camera showed material leaking from the sample head. A decision was made to stow the samples earlier than planned to prevent any further loss. On 28 October 2020 after safely moving the TAGSAM arm over to the open sample return capsule the head of the TAGSAM was successfully stowed. Pictures returned to Earth indicated that the head was successfully placed in the correct position and the sample return capsule was closed and ready for its journey back to Earth. OSIRIS-REx will stay at Bennu until March of 2021 when the return journey to Earth will begin. Arrival at Earth is planned for September 2023 when the sample return capsule will detach from the main spacecraft body and parachute to a landing in Utah in the USA. Only then will we know how much material has been returned, but it is a lot more than the minimum 60 grams. The scientists await!

Mission website can be found at **www.asteroidmission.org**

Hayabusa2 Returns Samples of Ryugu

The Japanese asteroid sample-return mission Hayabusa2 was launched toward the carbonaceous near Earth asteroid 162173 Ryugu on 3 December 2014, arriving at its destination on 27 June 2018. The spacecraft successfully explored the asteroid from orbit, and deposited three small rovers to explore the surface (a fourth rover failed before being released from the orbiter). The surface sampling was completed on 22 February 2019, and a small amount of regolith – or surface material – was successfully placed into a sample return capsule. On 12 November 2019 Hayabusa2 departed Ryugu for the return trip to Earth. Just over six years after launch the spacecraft jettisoned the sample return capsule, shortly after passing inside the orbit of the Moon. The onboard propulsion system performed a deflection manoeuvre so that the Hayabusa2 spacecraft would successfully pass the Earth and continue in orbit around the Sun for potential follow on fly-by missions to two other asteroids. The sample return capsule entered the Earths atmosphere just after 17.30hrs GMT on 5 December 2020. Following ejection of the heat shield a parachute was deployed which brought the capsule safely down in the Woomera area of South Australia

Artists impression of the Japanese Hayabusa2 spacecraft returning the sample return capsule to Earth. (JAXA/DLR German Aerospace Center)

shortly before dawn, and after a total flight of 5.24 billion kilometres. When daylight broke a recovery team successfully located the capsule and its precious cargo. Sample material will be shared with the Americans who will also share their samples of asteroid Bennu when their OSIRIS-Rex mission comes back to Earth in 2023.

Mission website can be found at **www.hayabusa2.jaxa.jp/en**

And Finally…

The Parker Solar Probe was launched on 12 August 2018 and continues its journey around the Sun. As of time of writing (late-2020) its orbit brings it to within 18 million kilometres of the Sun's surface. Continued gravity assists with Venus is reducing the perihelion or closest point to the Sun. By the end of mission in 2025 the perihelion distance will be down to 6.9 million kilometres. At that distance, the intense gravity of the Sun will have accelerated the probe to a staggering 690,000 kph.

Readers can follow the progress of the Parker Solar probe at the mission website: **parkersolarprobe.jhuapl.edu**

The successful JUNO mission to Jupiter is nearing its end. Arriving at Jupiter on 5 July 2016 is was expected to be in a 14 day orbit around Jupiter but issues with the onboard propulsion meant that JUNO remained in the initial 53 day orbit. This did not have a long-term impact on the science return. It just took longer. Mission extensions allowed the spacecraft to complete all the science goals. The mission has been granted a four year extension through to September 2025. This extension will allow some close flybys of the large

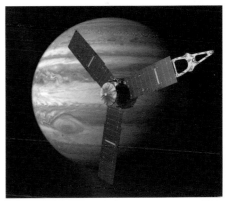

Artist's impression of the JUNO spacecraft in orbit around Jupiter. (NASA/JPL)

Galilean moons Ganymede, Europa, and Io. Mission details can be found here: **www.missionjuno.swri.edu**

As always, Solar System exploration continues to excite and inspire, and next year promises to be no different.

Anniversaries in 2022

Neil Haggath

Tycho's Supernova

450 years ago, a "new star" appeared in the constellation Cassiopeia, which briefly became bright enough to be visible in daylight. We now know that this was a supernova, one of eight visible to the naked eye in recorded history.

The most accurate observations of the phenomenon were made by the great Danish astronomer Tycho Brahe (1546–1601), and it became known as Tycho's Star. Tycho was perhaps the most meticulous observer of the pre-telescopic era; it's a tragic irony that he died just seven years before the invention of the telescope! The following year, he published *De nova et nullius aevi memoria prius visa stella* (Concerning the star, new and never before seen in the life or memory of anyone), with an analysis of his own observations and those of many others.

The star first reached naked eye visibility on or around 2 November 1572, and reached its peak brightness on 16 November, when it was recorded as about equal to Venus at its brightest. It gradually faded over many months, remaining visible to the naked eye until early 1574.

In those days, of course, even the intelligent still believed in astrology, and many, including Tycho himself, believed the "new star" to have some great astrological significance. In England, Queen Elizabeth I summoned an astrologer, though what he advised her is not known. In China, the young emperor was warned to consider his behaviour, as it was taken to be an evil omen. But perhaps the most bizarre interpretation was that of one Georg Busch, who claimed that the star was "formed by the ascension from Earth of human sins and wickedness", and would cause "all kinds of unpleasant phenomena, such as disease,

Portrait of Tycho Brahe by Austrian artist Eduard Ender. (Wikimedia Commons/Eduard Ender)

sudden death, bad weather and Frenchmen". As Sir Patrick Moore noted, "What the French thought about this is not on record!"[1]

In 1952, the radio source now designated 3C10 was discovered, and identified as the remnant of the 1572 supernova. Its optical counterpart, a faint nebula, was discovered in the 1960s, and an X-ray source in the 1970s by the Uhuru satellite. The latter is strangely designated Cepheus X-1, despite being in Cassiopeia.

From its light curve, deduced from the observations of Tycho and others, and studies of the remnant, we now know that the event was a Type 1a supernova, caused by the accretion of material from a companion star onto a white dwarf, triggering a gigantic thermonuclear explosion. Its distance is estimated as about 6,400 light years.

Historically, the great significance of Tycho's Star is that it challenged the ancient Aristotelian view, long held as religious dogma, that "the Heavens" were "perfect" and unchanging.

Sir William Herschel (1738–1822)

August sees the bicentenary of the death of the man who was perhaps the greatest visual observational astronomer who ever lived, Sir William Herschel.

Friedrich Wilhelm Herschel was born in Hanover – which was then united with Britain under the Royal House of Hanover – on 15 November 1738. However, he spent most of his life in England, and anglicised his given names to Frederick William.

As a teenager, he joined the Hanoverian Guards Regiment as a bandsman, but he left the army at the age of 19, and settled in England, where he became an accomplished musician and composer. In 1766, he settled in the fashionable spa town of Bath, where he became the organist of the Octagon Chapel.

Sir William Herschel, by English portrait painter Lemuel Francis Abbott, 1785. (Wikimedia Commons / Lemuel Francis Abbott)

In 1772, his sister Caroline Lucretia Herschel (1750–1848) joined him from Hanover and became his housekeeper, while also pursuing her own musical career. In 1780, William became Director of the Bath Orchestra, with Caroline often performing as soprano soloist.

---

1. Patrick Moore, *Explorers of Space*, Mitchell Beazley Publishers, 1986.

Had Herschel not been an astronomer, he would undoubtedly have had a lifelong career in music; he composed 24 symphonies and numerous concertos. But during the 1770s, his social status led to him mixing with various "philosophical gentlemen", and he acquired a keen interest in science, and especially astronomy.

He learned to grind speculum metal mirrors, and began building his own reflecting telescopes. He became an accomplished telescope maker, and established an international reputation; he sold at least 60 instruments to various British and European astronomers. While William was the skilled optician, he was assisted by his brother Alexander, who was a skilled mechanical craftsman. Legend has it that he would often become so engrossed in his work that Caroline had to feed him while he worked!

Due to the relatively poor optical qualities of speculum mirrors, Herschel invented a telescope design which eliminated the need for a secondary mirror. The primary mirror was slightly tilted, so that the focal point was at the top of the tube and at one side, where the image was viewed directly with an off-axis eyepiece. Such a design is still called a Herschelian reflector.

Herschel began observing in 1773, with a 6-inch reflector. Initially, his main interest was double stars, which at the time were thought to be only line of sight phenomena. It was thought that observations of them over time would show parallax shifts in their relative positions due to the Earth's orbital motion, enabling stellar distances to be measured. (It had long since been realised that nearby stars must show parallaxes with respect to more distant ones, but none had yet been measured, as the angles were too tiny to be measured with any instrument of the time.) He discovered over 800 pairs, and published two catalogues, which were presented to the Royal Society in 1782 and 1784.

Beginning in 1797, he measured many binaries again, and in 1802, correctly hypothesised that many of them were in fact gravitationally bound systems, with two stars orbiting each other.

On 13 March 1781, Herschel made the discovery which would make him famous. Still using his 6-inch reflector, he found what he initially thought was a comet, but soon realised was something else entirely. It was the planet now called Uranus – the first discovery of a planet in historic times. The discovery brought him acclaim throughout the astronomical world; he was elected a Fellow of the Royal Society, and awarded its prestigious Copley Medal.

The following year, King George III, who was an enthusiastic patron of the sciences, appointed Herschel as "King's Astronomer" (not be confused with Astronomer Royal), and awarded him a generous pension, which enabled him to devote himself full time to astronomy. William and Caroline moved to Datchet, near Windsor Castle, and later to Slough, where they would live for the remainder of William's life. Caroline, as William's assistant, was also given a salary by the King, making her, in effect, only the world's second female professional astronomer!

At Slough, Herschel built successively bigger telescopes, of 12 and then 18 inches aperture, with which to carry out a survey of deep sky objects. Unlike his contemporary Charles Messier (1730–1817), who compiled his famous catalogue of deep sky objects only to avoid mistaking them for comets, Herschel wanted to

Sir William Herschel's "Forty Foot" telescope as depicted in *Leisure Hour*, 2 November 1867, page 729. (Wikimedia Commons/Leisure Hour)

investigate their nature. He soon found that many "nebulae" were in fact clusters of stars. Between 1782 and 1802, he discovered an incredible 2,500 deep sky objects, details of which he published in three catalogues.

Throughout this work, he was always assisted by Caroline, who wrote down his observations as William described them to her verbally. She even compiled her own star catalogue, to identify reference stars and determine the positions of her brother's discoveries. Caroline also became an accomplished observer in her own right, discovering six comets and eleven deep sky objects of her own, using a telescope which William built for her.

Between 1785 and 1790, with a grant from the King, Herschel built his biggest telescope, with an aperture of 49 inches and a focal length of 40 feet. It was the biggest telescope in the world at the time – a distinction it would hold until 1845, when the Third Earl of Rosse built his 72-inch "Leviathan of Parsonstown". However, it proved cumbersome to use, and was not very successful; Herschel was not very satisfied with it, and most of his discoveries were made with his smaller instruments.

Herschel remained single until the age of 49, when he married a wealthy widow, Mary Baldwin. Their son John Frederick William (1792–1871) was born four years later.

John, of course, would himself go on to become a highly accomplished astronomer and scientific polymath. He extended his father's deep sky surveys to the southern sky, observing from South Africa, and discovered a further 1,754 objects of his own. In 1864, he combined his own discoveries with those of his father and aunt, in his *General Catalogue of Nebulae and Clusters of Stars*. 24 years later, this formed the basis of J. L. E. Dreyer's *New General Catalogue of Nebulae and Clusters of Stars*, or NGC, with which we are all familiar today. Dreyer added objects discovered by several other observers, but of the 7,840 objects in the NGC, a remarkable 4,265 were discovered by the Herschels!

In 1816, Herschel was awarded the honorary title of "Sir" by the Prince Regent, but not an official knighthood. In 1820, when the Astronomical Society of London – later to become the Royal Astronomical Society – was founded, he was elected as its first President. Until 2020, his "forty foot" telescope featured on the RAS's emblem.

Herschel's other discoveries were numerous. He discovered two satellites of Saturn, Mimas and Enceladus, and two of Uranus, Titania and Oberon. The prominent crater on Mimas is now named in his honour. He also discovered infrared radiation in 1800.

He studied the shape of the Galaxy, and deduced, almost correctly, that it has the form of a flat disc – though he wrongly believed the Sun to be at its centre. By studying the proper motions of stars, he deduced the direction of the Sun's motion and in 1783 determined the Solar Apex to within 10° of its correct position.

Sir William Herschel died at his home in Slough on 25 August 1822, aged 83. His and Caroline's former home in Bath, at 19 New King Street, now houses the Herschel Museum of Astronomy.

Harlow Shapley (1885–1972)

This year sees the 50th anniversary of the death of one of the pioneers of modern astrophysics – Harlow Shapley, the man who first measured the size of our Galaxy.

Shapley was born on 2 November 1885 in Nashville, Missouri. His route to becoming as astronomer was somewhat convoluted. He dropped out of school at 16, with only a basic education, to become a newspaper reporter. He later returned to school and completed his high school education.

In 1907, he went to the University of Missouri, intending to enrol in its new School of Journalism – only to find that its opening had been delayed by a year. Rather than wait, he decided to study something else instead.

Harlow Shapley. (John McCue)

The first subject in the course directory was archaeology, but he couldn't pronounce it – so he settled on the next one, astronomy!

For a man who embarked on his chosen career by accident, Shapley soon proved remarkably proficient in it. After graduating, he gained a postgraduate fellowship to Princeton, where he gained his Ph.D. under Henry Norris Russell (1877–1957), the "father of modern astrophysics". He was one of the first to realise that Cepheid variable stars are pulsating stars, and used their period-luminosity relationship, discovered by Henrietta Swan Leavitt (1868–1921) in 1912, to determine the distances of globular clusters, thereby establishing that they form a halo around the Galaxy. He also used RR Lyrae variables, which were also found to have a period-luminosity relationship, to determine the size and shape of our Galaxy – which he found to be far larger than previously believed – and the Sun's position in it.

However, Shapley believed that our Galaxy was the only one, and that the so-called "spiral nebulae" were small-scale structures within it. In 1920, while working at Mount Wilson Observatory, he took part in the famous "Great Debate" with Heber Doust Curtis (1872–1942), who advocated the view that they were separate galaxies. The Debate is described in detail in my article in the *Yearbook of Astronomy 2020*.

The Debate was inconclusive, but three years later, Edwin Hubble (1889–1953) settled the matter, when he determined the distance of the Andromeda Galaxy, also using Cepheids.

In the manner of a true scientist, Shapley graciously accepted that he had been wrong, and later studied the distribution of galaxies, mapping 76,000 of them between 1925 and 1932. He was one of the first to realise that galaxies form clusters, and discovered what is now named the Shapley Supercluster.

In 1921, Shapley became Director of Harvard College Observatory, a post which he held until he retired in 1952. In the 1940s, he helped to found the National Science Foundation, and in 1947, was elected President of the American Association for the Advancement of Science.

In 1953, Shapley proposed the concept of a "liquid water belt" around stars – that is, what we now call a habitable zone or Goldilocks zone.

Harlow Shapley died in Boulder, Colorado, on 20 October 1972, aged 86.

Pioneer 10

Another 50th anniversary this year is that of the historic space probe Pioneer 10. This is described in my article *A True Pioneer of Planetary Exploration*, in the March monthly notes.

Apollo 17

50 years ago in December, what was arguably humanity's greatest adventure came to an end, with Apollo 17, the last Moon landing mission.

Apollo 17 was launched on 7 December 1972. Its Commander was Captain Eugene A. "Gene" Cernan USN, aged 38, who was making his third spaceflight, having previously flown on Gemini 9 and Apollo 10; he became one of only three men to fly to the Moon twice. His companions were both first-timers – Commander Ronald E. Evans USN, 39, was the Command Module Pilot, and Harrison H. "Jack" Schmitt, 37, the Lunar Module Pilot. Their Command and Service Module was patriotically named *America*, and the Lunar Module *Challenger*.

Schmitt, uniquely among the astronauts, had a Ph.D. in geology. He was one of a group of scientist-astronauts selected in 1965; he had originally been assigned to Apollo 18, and had instructed the other astronauts in geology. When the last three planned missions were cancelled due to budget cuts, NASA didn't want to lose the opportunity to send a geologist to the Moon, so Schmitt was transferred to Apollo 17.

Cernan and Schmitt landed *Challenger* in the Taurus-Littrow Valley, a highland region near the edge of Mare Serenitatis. They stayed on the Moon for 75 hours, the longest of any mission, made three EVAs, each lasting over seven hours, and drove a total of 34 km in their Lunar Rover. They collected a record 110 kg of rock samples.

During their second EVA, they discovered some unusual orange soil, which turned out to consist of tiny beads of volcanic glass.

They also kept TV audiences entertained with their humour; during one EVA, Cernan sang: "I was strolling on the Moon one day…"!

For the only time in the programme, Apollo 17 carried five mice, in a biological experiment to study the effects of cosmic rays on living organisms. One died during the return journey, but four survived to return to Earth.

The crew returned to Earth on 19 December, after a flight time of 12 days 13 hours 52 minutes.

As the Commander always left the Lunar Module first and returned to it last, Gene Cernan was the last person, so far, to walk on the surface of the Moon. As he left it for the last time, his parting words evoked the message left there by Apollo 11: "We leave as we came, … with peace and hope for all mankind."

Apollo 17 crew: Gene Cernan (front), Jack Schmitt (rear left) and Ron Evans. (Wikimedia Commons/NASA)

The Yearbook of Astronomy Cover Image 1962 and 2022

Steve Brown

It has been sixty years since the first publication of this Yearbook, and astronomy was very different back in 1962. Observation for the amateur was very much still a visual endeavour, with photography just beginning to make its mark. Things are very different today, with equipment and technology available that even the professionals of six decades ago could barely dream of. Time moves on, things change and progress is made. It is comforting to know however, that some things will always be there, like our closest neighbour in space, the Moon. There might be human boot prints on the lunar surface, something not yet achieved when the first *Yearbook of Astronomy* was published, and professional astronomers might have mapped its surface in exquisite detail, but these things do not diminish the beauty and romance of our only natural satellite. Many an amateur astronomer has honed their skills observing and photographing the Moon and when we observe, sketch or photograph the Moon, we are making a connection down through history to a long line of people who have done the same. From the days of early telescopic observation and drawings, through to today's high definition images, we are all adding to a rich history of lunar study.

Just as the Moon has been a constant companion of the astronomer, for the past sixty years so too has the *Yearbook of Astronomy*. The photograph on the front cover is the same

Yearbook of Astronomy 1962. (Published by Eyre & Spottiswoode Publishers Ltd)

Francis Gladheim Pease. (Wikimedia Commons)

as that of the very first edition back in 1962. Pictured is the south-central portion of the third quarter Moon at around 60 per cent illumination. The craters Tycho and Clavius are well shown, with the curving chain of craters within Clavius particularly well captured. The impact rays from Tycho are also clearly visible emanating from the crater and are particularly noticeable on the darker plain of Mare Nubium to the bottom right of the picture. The photograph was taken on 15 September 1919 by Francis Gladheim Pease through the 100-inch Hooker telescope at Mount Wilson Observatory, California. At the time, this was the largest aperture telescope in the world. The next time you go out to observe the Moon, take a moment to think about all those before you who have done the same. All those lives lived out down here on Earth, while up above resides the Moon, our eternal neighbour in space. We might face trials and tribulations down here on Earth, but the Moon will always be there, just waiting for us to look up and take a moment to allow a little awe and wonder into our daily routine.

The 100-inch Hooker telescope through which the original image was taken. (Wikimedia Commons / Ken Spencer)

A History of the Amateur Astronomical Society
1962 to 2022

Allan Chapman

The amateur astronomical society was a Victorian creation, and hugely influenced – especially in the English-speaking world – from the 1850s onwards by the example and writings of that intellectual ancestor of Sir Patrick Moore, the Revd Thomas William Webb, whose popularisation of the new DIY silver-on-glass reflecting telescopes and garden observatory buildings first got people looking sky-ward in a serious way. Leeds, Liverpool, Cardiff and elsewhere all had thriving astronomical societies, complete with active lady members, by the time of the founding of the BAA in 1890. The membership of these societies, however, was dominated

Lancaster University Astronomical Society members observing the transit of Mercury, 9 May 1970, from the Hazelrigg Field Station, Lancaster. Left to right: Christopher J. V. Hunt, Allan Chapman, Jim May. (Allan Chapman)

by urban middle-class professionals, such as GPs, teachers, solicitors, clergymen and businessmen, most of whom lived in the better-off parts of their cities, with scarcely a working man or woman from the industrial districts.

Yet in the same way that astronomy itself changed profoundly between 1890 and 1962, so too did British society. Two World Wars, financial turmoil, the Cold War, and massive improvements in public education had turned the Victorian world upside down. Working people now enjoyed living standards unimaginable to their grandparents, as cheap cars, longer holidays, and foreign holidays had transformed leisure time, while overcrowded gas-lit cottages had been replaced by luxurious council houses. For by 1962, after passing the right exams, the son or daughter of a coal miner might rise up to become a brain surgeon, a barrister, or a professor of astrophysics.

And then there were the 'New Physics' of relativity, galactic infinity, radio astronomy and the 'Space Race', while radio and television and the communications revolution had brought a whole realm of knowledge straight into everyone's living room. Likewise, the Russian artificial satellite Sputnik after 1957 got everyone out into their back gardens to see the man-made star fly over, and everyone with access to an ex-Royal-Navy telescope was scrutinizing the full moon hoping to see a Russian rocket – or a Little Green Man! For it was all these new things celestial during the 1950s and 1960s that inspired people to start meeting informally together, often in pubs, to discuss the new astronomical fascination.

And these groups included a whole cross-section of society: a vicar, a science teacher, a bricklayer, an office clerk and a garage mechanic, along with some bright youngsters – provided, that is, the pub landlord would let them in. Some would relate their efforts to grind and polish a 6-inch mirror, others would be struggling with Einstein, while there were invariably those 'weirdos' who had met Martians or were into horoscopes.

All of the above is historical truth, as I know from personal experience in my own youth, and what a colourful, if often exhausting, bunch they could be. I recall, for example, 'The Android', with his flow of physics gibberish and personal acquaintance with superior aliens who wanted to save us poor benighted humans from ourselves. Unfortunately, these wise folks never suggested that he should wash and change his clothes, for I am sure that they could smell him on Mars, Alpha Centauri, or wherever they came from! Others were into 'galactic radiation astrology', anti-gravity, and US Pentagon conspiracy cover-ups regarding flying saucers, or secret aliens who were already among us on earth. Indeed, the new coming together of genuine astronomical enthusiasts also drew in a colourful diaspora of 'independent thinkers'.

Salford Astronomical Society Observatory, Chaseley Fields, Salford, housing a professionally-made 18-inch Newtonian reflector on a fork mount which was once part of Manchester University's original equipment at Jodrell Bank. Professor Zdeněk Kopal, of Manchester University, very kindly arranged for this now redundant optical telescope to be given to the Salford Society around 1970. Salford City Council generously built the Observatory building. (Photographer unknown, Salford Astronomical Society archive picture)

Yet all these colourful folks apart, most astronomical groups were made up of intelligent men and women who lived in the real world, and had become captivated by the real science of the sky. The society of which I have been a member ever since its formation in the mid-1960s is that of Salford, near Manchester, which I now have the honour to serve as President and Christmas Lecturer. I recall my father seeing a notice in a local paper that one Arthur Taylor, a master carpenter and builder working for the City Council, and Alan Whittaker, a retired bank manager, were inviting people to meet up and share their astronomical passion. This would quickly crystallise into the Salford Astronomical Society. The early pub meetings drew in a fair crowd, and before long we had been given a room in the Buile Hill Park Museum. Manchester had an Astronomical Society going back to 1903, of which Alan Whittaker was a long-standing member, and the two societies have always worked closely together.

Over the ensuing decades, I came to speak to scores of local societies across the length and breadth of Great Britain. Some, such as Leeds, Liverpool, Cardiff, and Manchester, were the venerable founding fathers and mothers of amateur astronomy, while a vastly greater number trace their origins to the 1960s: the 'Space Race' societies. And being a great lover of jokes and funny tales, as well as a patriotic 'Red Rose' Lancastrian, I enjoy having fun when speaking to 'White Rose' Yorkshire, societies. In particular, when I am with old friends in the Mexborough and Swinton Society, West Riding of Yorkshire, I will say 'I see that at last civilisation is spreading to Yorkshire, for you have elected a Lancastrian [me] as your President' – followed by humorous growls and snarls from the floor!

Jokes, 'space ships', and educational opportunities apart, however, one fundamental aspect of space age just like Victorian astronomy was access to high-quality instrumentation, for telescopes have always been the golden key to unlocking the glories of the sky. Most of the early societies had active telescope-making groups. People prided themselves on being able to figure a fine 6-inch mirror: always regarded as the minimum aperture with which to do 'serious' amateur astronomy, and serious glass-workers won especial admiration for their 8- and 10-inch masterpieces. Most societies also aspired to build an observatory with a larger-aperture instrument to be shared by all society members. After all, these new societies contained, and still contain, members possessing a diversity of

Mexborough and Swinton Astronomical Society open-day event in 2010. Mike Collinson (hatless) demonstrates a telescope. (Mexborough and Swinton Astronomical Society archive photo, 2010)

skills, such as building, bricklaying, and light engineering, for amateur astronomers tend to be practical people.

Then when in 1970 one had figured a fine mirror which had passed the iconic Foucault Test, enormous ingenuity could be exercised in mounting it, be it for a private or a society observatory. Miscellaneous pipes, tubes, and plumbing fittings would be adapted to form an equatorial mount, while an old car axle, rescued from a local scrap yard, could be adapted to provide a smooth-turning equatorial head. (And I know that all this is true, for from my childhood onwards I did this myself, and have never been able to resist the allure of a junk shop, hardware store, or scrap yard. And I still have some of my early creations to prove it!)

Lancaster University Astronomical Society's astronomical observatory was designed and built during the summer vacation of 1970 by Allan Chapman, the Society's founder. The 6-foot diameter marine-quality plywood dome housed a Japanese 3-inch aperture equatorial refractor. The Observatory, with timber walls and dome, was constructed in a space kindly provided by the University Engineering Sciences Department, and then transferred to Bailrigg Hill, on University land. The dome was balanced upon six kitchen ceiling washing-line pulley wheels purchased from a local hardware shop, rotating, very smoothly, around a circular steel track. The dome also had a sliding roof shutter. When, some years later, the Observatory was replaced by a permanent professionally-built roof-top building with a larger telescope in the University Physical Sciences Department, the original Observatory was sold to a Lancaster amateur astronomer. (Allan Chapman)

Then if space in the back garden permitted, one's efforts were truly crowned if the telescope could be permanently mounted in a run-off shed, or even a domed observatory. Garden sheds resounded with construction noises, as DIY enthusiasts set about building their own mini versions of Mount Palomar between the cabbage patch and the dahlias. I also hand-built a six-foot rotating observatory dome with a sliding roof slit from marine-standard plywood for the student Astronomical Society over the summer of 1970, when an undergraduate at Lancaster University. Sometimes, however, non-astronomical neighbours complained, believing these exotic creations were to be used by Peeping Toms, or even by Russian secret agents. This was, after all, the height of the Cold War, WWII had been only twenty years ago, and James Bond's heroics were filling the cinemas.

But it was really our old WWII enemy, Japan that came to the aid of many aspiring astronomers, as the ingenious Japanese capitalised on their old ex-Zeiss German optical technology to flood the market with relatively cheap, high-quality, ready to use optics. I still have my c.1966 60 mm aperture Japanese refractor, and a 4.5-inch aperture Newtonian reflector, both fully mounted, which brought the moon so close that you felt you could touch it. And both were bought at a price well below that of German equivalents.

All these instruments would only give modest magnifications by modern standards, and much amateur astronomy was devoted to the moon and planets. What were still termed 'nebulae' remained little more than faint smudges, especially when viewed through the coal-smoke-polluted skies of the still steam-powered Britain of 1966.

But there was always photography, and like countless amateurs I fiddled about with trying to adapt an old box camera to a telescope eyepiece, or slightly later a cheap Japanese single lens 35mm roll film reflex camera. Photographs of nebulous smudges, hazy planets, and lunar craters won local celebrity status for serious amateur astronomers in the late 1960s, as they were proudly handed around society meetings.

The waxing moon, photographed with the 3-inch aperture telescope in the Lancaster University Astronomical Society observatory, c.1971, on a 35mm b/w negative. (Allan Chapman)

And then came the seismic shift brought about by the digital revolution, and the mere couple of decades that separated the age of the green-screened TV-sized desk computers from that of the smartphone propelled amateur astronomy out of what, in retrospect, seems like the Jurassic into the modern age. Then with the internet and broadband, the tectonic plates shifted yet further.

We must not forget, moreover, that digitalisation and the internet transformed not just science, but also the whole way in which we lived and worked. Heavy industrial manufacture needed fewer and fewer human hands, as cars, washing machines, and even precision optical devices were being mass-manufactured by computer-controlled robots. In 1962, one made the daily trek to the factory or office for eight or more hours' grind, and often had only about three weeks' holiday per year. But the new robotics released millions of human hands, and while this caused severe economic distress at first as old labour-intensive industries either vanished or became computerised, countless millions learned new skills, worked 'flexitime', became self-employed or worked at home from a keyboard. Few amateur astronomers could manage to stay up observing until 4 a.m. and still walk through the factory gates at 8.00 a.m. in 1962, but by 2000 many could adjust their working day so that astronomy, family time, and work could fruitfully co-exist together. This released a vast amount of potential observing time.

Digitalisation and the internet also had the power to make amateur astronomy global – just as the professionals were doing. British, European, American, Australasian and Far Eastern amateurs were now in contact with each other in a way that no one could have imagined in the postage stamp days of 1965. News of a sudden solar flare or a distant galactic supernova eruption could now be transmitted around the planet 'in the twinkling of an eye', and telescopes and video cameras from Tokyo to Taunton directed towards it. And the instruments being employed by these new, international amateur astronomers were not the home-made wonders of DIY ingenuity, but exquisite 12- or 14-inch Schmidt Cassegrain telescopes, harvesting and storing their data through sophisticated CCD cameras and software processing systems - an international synchronicity of instrumentation, no less. Yet all of this high-quality optical and computer kit, instead of costing millions, need not cost much more than a good used car.

Amateur astronomical photography was no longer a matter of dodging clouds and later admiring fuzzy images of Saturn's rings or the lunar crater Copernicus, so much as using a whole armoury of digital techniques, such as multiple stacked images, which made it possible to optimise fine detail while discarding blurs caused by transient atmospheric turbulences. Indeed, these new relatively cheap high-quality telescopes and ancillary CCD equipment perhaps did more than anything else to close up the old amateur–professional gulf. For by 2010, it was possible for a serious amateur astronomer who earned his living as a bus driver in Preston to obtain better images of deep-sky objects than would have been possible with the mighty 200-inch Mt Palomar telescope and glass plates in 1950. And if our Preston amateur took his lightweight portable gear on holiday to Tenerife, and stacked yet more CCD images well up the mountain, he could get even better ones. Indeed, in some ways only the Hubble and other space telescopes can now claim an undisputed lead in astronomical photography – and I even recall conversations about a planned 'Humble Space Telescope', to be launched using commercial rocketry, and owned and controlled by amateurs.

This home-made 4.5-inch aperture refractor, built by the author, c.1966, had a 5-foot focal length single element stopped-down object glass, purchased from a local optician and conjoined with an old microscope eyepiece. The rack and pinion focussing mechanism was adapted from a broken Victorian magic lantern. A former plastic water pipe, rescued from a scrap heap, became the tube, while the object glass mounting cell was fashioned from an old paint tin. The wooden tripod, mount, and fittings were all designed and built by the author. True junk-yard astronomy, but it gave decent images of the moon. (Allan Chapman)

Of course, the advent of the 'go to' telescopes made the location of faint galactic objects quick and easy. But committed astrophotographers, clever with their hands and digitally savvy, could go even further, building fully-automated home observatories. All that the astronomer now needed to do was program the computer control system, then go to bed. At the appropriate time of night, if the sensors signalled a clear sky, the observatory roof would open automatically, and the 'go to' telescope lock onto the appropriate galaxy, variable star, or supernova currently being monitored, and then run a series of images. When the task was complete, the observatory would automatically close up and shut down.

Then in the morning, the astronomer would check his or her computer and examine at leisure the run of pictures taken the night before. The serious researcher would then process the images, and send them to whichever professional or amateur body was coordinating the project in which they were taking part, to add to the globally coordinated body of new astronomical data.

And while, no doubt, our electronic astronomer is a member of a local astronomical society, such a person would automatically be a member of an international 'society' of observers and researchers who work together, be they in Preston, Beijing, or Portland, Oregon. And that 'society' will most likely contain bus drivers, surgeons, teachers, and professors, brought 'virtually' together in their pursuit of supernovae, variable stars, or exoplanets.

In one respect, however, astronomical societies have not changed between 1962 and 2022, and that is in their global commitment to 'outreach'. Whether this involves giving astronomy talks in schools, 'sidewalk astronomy' showing the heavens to passers- by in the street, or taking the stand at a local science festival, I have never ceased to be impressed by the commitment of amateur astronomers in sharing the celestial wonders with others. And whether your own society still meets in a back room at 'The Wagon and Horses', a church hall, or a local civic centre, or whether it is 'virtual' and international, that passion to delight in and understand the night sky, combined with a desire to share it with others, is just as strong in 2022 as it was back in 1962.

Tom Fearn demonstrates a telescope to Salford Astronomical Society members, c.1967. (Salford Astronomical Society archive photograph)

Expanding Cosmic Horizons

Martin Rees

The history of astronomy has been a story of expanding horizons. I think of it as a series of Copernican revolutions. The first was of course the revolution inspired by Copernicus himself – dethroning the Earth from its central position and grasping the basic structure of the solar system. Then the second revolution – realizing that our Sun was just one star in the Milky Way. Finally the third – the recognition a century ago that our Milky Way was itself just one galaxy amongst billions. We have understood the nature of stars and galaxies, and that everything they are made of emerged from a 'big bang'. And recent ideas – still speculative – suggest that we may be due for a fourth Copernican revolution: that 'our' big bang may be one of many, and that we actually live in a 'multiverse'.

But as I will explain, 'observational cosmology' as a real science did not really start until the 1960s. Like the *Yearbook of Astronomy*, it is celebrating its 'diamond jubilee'

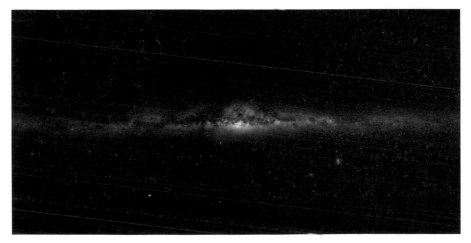

This 360-degree panoramic image shows the stars of our Milky Way. These are concentrated in a thin disc in which we are embedded. The view is 'patchy' because there are obscuring dust clouds along some lines of sight, in particular, the central region, where there lurks a black hole with a mass of four million suns, is best observed in the infra red and radio band. (ESO/S. Brunier)

Widening Horizons: The Extragalactic Realm

But let's start with a flashback to earlier eras. Eighteenth- and nineteenth-century astronomers interpreted the Milky Way as a flattened disc of stars in which our Sun was embedded. But they had no appreciation that anything lay beyond: this stellar system was, to them, 'the universe'. But horizons hugely broadened in the 1920s.

In the Great Debate[1] held in Washington, DC around 100 years ago Harlow Shapley maintained that our Milky Way was pre-eminent and all the 'nebulae' were really part of it. In contrast, Heber Curtis argued that some of the fuzzy objects were separate galaxies – fully the equal of our entire Milky Way. In particular, the object then known as the Andromeda Nebula was recognised as a separate system comparable in size and mass to our own Milky Way – a spinning disc, viewed obliquely, containing at least 200 billion stars, orbiting around a central hub. It is 2.5 million light years from us. Viewed by an alien astronomer from that distance, our Milky Way would look similar (albeit slightly smaller).

We now know that galaxies, with different morphologies, are the dominant

The Andromeda Galaxy, a spinning disc viewed obliquely, contains more than 2 billion stars orbiting around a central hub. It was only in the1920s that this system was recognised as a separate galaxy, fully equal to our Milky Way. (Wikimedia Commons / Adam Evans)

features of the large-sale cosmic scene; and that they are not scattered randomly, but concentrated in groups and clusters. The next big discovery, dating from the

1930s, was that the light from these galaxies displayed a 'red shift'. If interpreted as a Doppler effect this implied that were receding from each other – indeed the further away they were, the faster their recession seemed to be.

Back in the 1920s Alexander Friedmann and Georges Lemaître had found solutions of Einstein's equations that described expanding universes. And in 1948 Fred Hoyle, Hermann Bondi and Thomas Gold offered a contrasting view that the universe was in a "steady state": as space expanded and galaxies dispersed, new ones forming 'in the gaps' so that the average properties of the universe were unchanging.

But until the 1960s it was not possible to discriminate among different models. Telescopes did not have the power to probe deep enough into space – and hence far enough back in time – to detect the changes in the expansion rate, or in the mean properties of galaxies, which might have been expected to occur as the universe aged.

A breakthrough on this front came in 1963 with the discovery of quasars – hyper luminous beacons in the centres of some galaxies, compact enough to vary within hours or days, but which vastly outshine their host galaxy. This discovery revealed that galaxies contained something more than stars and gas. That 'something' is a huge black hole lurking in their centres, though it took a further decade before that was properly appreciated. Quasars are especially

An artist's impression showing ULAS J1120+0641, the most distant quasar yet found. A jet is seen emerging from a black hole millions of times the mass of the sun, embedded in gas of a newly-forming galaxy. (ESO/M. Kornmesser)

bright because they are energised by emission from magnetized gas swirling into a central black hole. This process can generate the jets that inflate the giant radio-emitting lobes surrounding some galaxies. And because they are so bright they allowed optical and radio telescopes, for the first time, to probe back to an era when conditions were indeed different – and to find that when galaxies were young their centres were typically far more 'active' than they are today. This was the death-knell of the 'steady state theory'.

Among the huge range of images generated by the Hubble Space Telescope, this showing the 'Hubble Ultra-Deep Field' is one of the most remarkable. It was obtained by pointing the telescope at the same small patch of sky for several days; many of the galaxies in the picture are more than 10 billion light years away and are being seen as they were when they were recently-formed. (NASA/ESA/H. Teplitz and M. Rafelski (IPAC/Caltech)/A. Koekemoer (STScI)/R. Windhorst (Arizona State University)/Z. Levay (STScI)

Galaxies – some disc-like, resembling our Milky Way or Andromeda; others amorphous 'ellipticals', and with a broad spread in their sizes – are the basic constituents of our expanding universe. And we know that they are held together, and stopped from flying apart, not just by the gravity of the stars and gas we see, but by an invisible swarm of particles – the dark matter – whose total mass is about five times larger than the visible constituents. And most galaxies harbour massive black holes in their centres – relics of quasar-like activity when they were young.

But how much can we actually understand about galaxies themselves? Physicists who study particles can probe them, and crash them together in accelerators. Astronomers plainly can not crash real galaxies together. And the structures of galaxies change so slowly that in a human lifetime we only see a snapshot of each. But we can do experiments in a speeded-up 'virtual universe' – computer simulations, incorporating gravity and gas dynamics.

We can redo such simulations, making different assumptions about the mass of stars, gas and dark matter in each galaxy, and so forth, to see which matches the data best.

We can test ideas on how galaxies evolve by observing eras when they were young. Unlike geologists, we can actually observe the past, rather than just seeking fossils. The Hubble Space Telescope has been used to study 'deep fields', each encompassing a patch of sky just a few arc minutes across. In such a tiny patch you can see hundreds of smudges. These are galaxies, some comparable to our own, but so far away that their light set out more than 10 billion years ago – they are being viewed when they were newly formed. And thanks to much broader surveys using ground-based telescopes, the extent to which galaxies are grouped into clusters – an important clue to how they formed – is now well documented.

The Remote Past

But what took place still further back in time, before there were galaxies? The key evidence here, dating back to Arno Penzias and Robert Wilson during the 1960s, is that intergalactic space is not completely cold. It is warmed to nearly 3 degrees above absolute zero by weak microwaves, known to have an almost exact black body spectrum. This is the 'afterglow of creation' – the cooled and diluted relic of an era when everything was squeezed hot and dense – and is one of several lines of evidence that have allowed us to firm up the 'hot big bang' scenario.

In the hot early phases, the universe was opaque, rather like the inside of a star. But when the temperature fell to around 3,000 K, the electrons combined with nuclei to create neutral atoms. This was after about 300,000 years of expansion. Measurements of the background radiation provide a direct picture of that

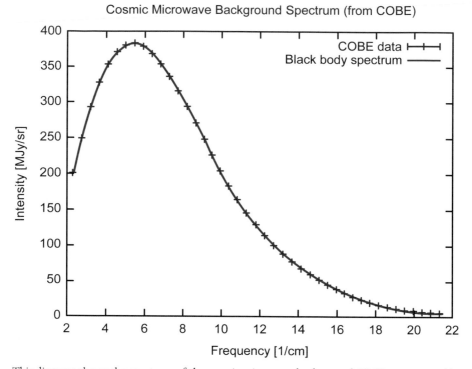

Cosmic Microwave Background Spectrum (from COBE)

This diagram shows the spectrum of the cosmic microwave background (CMB) as measured by NASA's Cosmic Background Explorer (COBE) spacecraft. This is the cooled and diluted relic of the universe's hot dense beginning. The measurement is precise (the uncertainty being no larger than the thickness of the line on the graph). The spectrum is almost exactly that of a 'black body' that has come into equilibrium with its surroundings. (Wikimedia Commons/NASA/COBE)

pre-galactic era, because neutral atoms, unlike free electrons, cannot scatter the radiation.

Astronomers can confidently trace back to one second. The temperature would then be about 10 billion degrees C; nuclear reactions would create a mix of hydrogen, helium and deuterium, and the calculated proportions agree gratifyingly with what is actually observed. Indeed we can be fairly confident in extrapolating back to a nanosecond. That is when each particle had about as much energy as can be achieved in a particle accelerator like the one at CERN in Geneva; and the entire visible universe was squeezed to the size of our solar system.

The Ultra-Early Universe and the Far Future

But questions like "Why the early universe contained the actual mix we observe of protons, photons and dark matter?" and 'Why is it expanding the way it is?' take us back to the even briefer instants when our universe was hugely more compressed still. And a warning: laboratory experiments are unable to simulate the conditions that prevailed at these ultra-early eras, so our theories are on shakier ground.

According to a popular theory devised around 40 years ago, the entire volume we can see with our telescopes 'inflated' to a hyper-dense blob the size of an apple in about a trillionth of a trillionth of a trillionth of a second – and the slightly 'over dense' regions that eventually condensed into galaxies, were generated by microscopic quantum fluctuations in that era. The generic idea of an 'inflationary phase' is taken seriously, because it can explain some otherwise-puzzling features of our universe. But because we do not know the physics that prevailed at the time, the details remain speculative.

In 1998, cosmologists had a big surprise, which did not change anything I have said about cosmic history, but did alter our perspective on the far future. It was expected that the cosmic expansion was slowing down because of the gravitational pull that galaxies exert on each other. But the red shift versus distance diagram of a particular type of supernovae (type 1a) famously revealed that the expansion was slower five billion years ago than it is today: it is now accelerating. Gravitational attraction was seemingly overwhelmed by a mysterious new force latent in empty space that pushes galaxies away from each other.

Long-range forecasts are seldom reliable, but the best and most 'conservative' bet is that we have almost an eternity ahead – an ever colder and ever emptier cosmos. Galaxies accelerate away and disappear over an 'event horizon', rather like an inside out version of what happens when things fall into a black hole. All that will remain in view are the remnants of our Galaxy, the Andromeda Galaxy and smaller neighbours. Protons may decay, dark matter particles annihilate, occasional flashes when black holes evaporate – and then darkness.

How Big is Physical Reality?

Now another couple of basic questions: How much space is there altogether? How large is physical reality?

However powerful our telescopes are, we can only see a finite volume – a finite number of galaxies. That is essentially because there is a horizon – a shell around us, delineating the distance light can have travelled (across expanding space) since the big bang. But that shell has no more physical significance than the circle that delineates your horizon if you are in the middle of the ocean. Unless we take a very non-

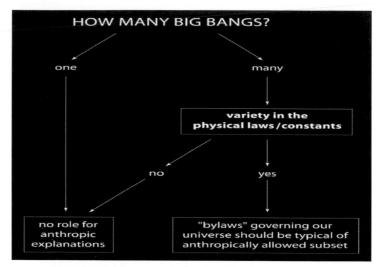

Whether we are part of a multiverse – whether 'our' big bang was unique, or is part of an ensemble – is still a matter for conjecture. We must aim to decide which branch of this 'decision tree' reflects reality. If there are many big bangs, which cool down differently so that their aftermaths are governed by different laws, then we will find ourselves not in a 'typical' universe, but one where the prevailing physics is conducive to the emergence of complexity and life. This is called 'anthropic selection'. (Martin Rees)

Copernican view, and imagine that we are at the centre of a spherical cosmos which has an edge just beyond our horizon, there are galaxies beyond the horizon which we can never (even in principle) observe if the expansion continues to accelerate.

We of course have no idea how far this unobservable domain extends. If it stretched far enough, then all combinatorial possibilities could be repeated. Far beyond the horizon, we could all have avatars. But even conservative astronomers are confident that the volume of space-time within range of our telescopes is only a tiny fraction of the aftermath of our big bang.

And there is something else. Some versions of 'inflation' – for instance a model developed by the Russian cosmologist Andrei Linde – suggest that 'our' big bang isn't the only one: it could be just one island of space-time in a vast archipelago.

A challenge for twenty first-century physics is to answer two questions. First, are there many 'big bangs' rather than just one? Second, if there are many, are they all governed by the same physics or not? But some theorists speculate that space itself – the vacuum in which these laws are played out – could have different 'phases'. If the aftermath of different 'big bangs' led to different vacua, they would evolve differently. What we call ' laws of nature' may in this grander perspective be local bylaws governing our cosmic patch.

The 39 metre diameter Extremely Large Telescope (ELT), currently under construction on Cerro Armazones in Chile's Atacama Desert, will be the largest optical/infrared telescope in the world, and will tackle some of the biggest scientific challenges of our time. The ability of the ELT to detect exceedingly faint objects will have a transformative influence on our understanding of many phenomena – from Earth like planets orbiting nearby stars to ultra-distant galaxies whose light set out when they were newly formed. (ESO/L. Calçada)

If there is just one big bang, then we would aspire to pin down why the numbers describing our Universe have the values we measure (the masses of protons, electrons, etc, and the forces between them). But if there are many big bangs, then physical reality is hugely grander than we had have traditionally envisioned. Many of the universes would be sterile or stillborn: for instance, gravity could be so strong that stars and planets would be too small and short-lived to allow any complexity to emerge in them. Or there may be no stable elements other than hydrogen and therefore no chemistry. So 'our' universe would not be a typical one: it would be among the subset where the laws were propitious for emergent complexity.

Some do not like the multiverse concept; it means that we will never have neat explanations for the fundamental numbers, which may in this grander perspective be just environmental accidents. This naturally disappoints ambitious theorists. But our preferences are irrelevant to the way physical reality actually is – so we should surely be open-minded.

Indeed, there is an intellectual and aesthetic upside. If we are in a multiverse, it would imply a fourth – and grandest – Copernican revolution; we have had the Copernican revolution itself, then the realization that there are billions of planetary systems in our galaxy, then that there are billions of galaxies in our observable universe. But we would then realize that not merely is our observable domain a tiny fraction of the aftermath of our big bang, but our big bang is part of an infinite and unimaginably diverse ensemble.

Some years ago I was on a panel at Stanford University where we were asked by someone in the audience how much we would bet on the multiverse concept. I said that on the scale 'Would you bet your goldfish, your dog, or your life?' I was nearly at the dog level. Andrei Linde, who had spent 25 years promoting 'eternal inflation', said he would almost bet his life. Later, on being told this, the great theorist Steven Weinberg said that he would happily bet Martin Rees's dog and Andrei Linde's life.

Andrei Linde, my dog, and I will all be dead before this is settled. But none of this should be dismissed as metaphysics. It is speculative science – exciting science. And it may be true.

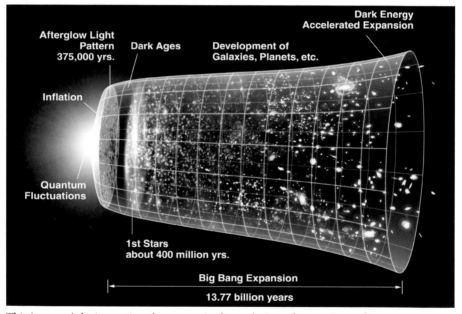

This 'cartoon' depicts various key stages in the evolution of our universe from its mysterious beginning in a 'big bang'. For the first microsecond everything would be so hot and dense that experimental physics offers few clues. But we have evidence about the first few seconds from the present abundance of helium and deuterium, mainly created by nuclear processes at that era. When the afterglow of the big bang cooled below about 3000 K the universe became dark, and remained so until the first stars formed and lit it up again. (Wikimedia Commons / NASA / WMAP Science Team)

Concluding Thoughts

Now we can, with confidence, trace cosmic history back 13.8 billion years, to a moment only a billionth of a second after the Big Bang. Astronomers have now pinned down our universe's expansion rate, the mean density of its main constituents, and other key numbers, to a precision of one or two percent. We are learning how cosmic structures – galaxies and clusters – emerged and evolved from amorphous dense beginnings. When the history of science is written, that amazing progress will be acclaimed as one of its greatest triumphs. Other fields have opened up too. Exoplanet research (which I have had no space to address here) is only 20 years old, and serious work in astrobiology is really only starting. Today's young astronomers will have an exciting 60 years ahead of them!

But progress will, as in the past 60 years, depend on ever-improving instruments – both on the ground and in space – and on bright ideas from a new generation of astronomers. What Edwin Hubble wrote in the 1930s remains a good maxim today: "Only when empirical resources are exhausted should we enter the dreamy realm of speculation."

The astronomer Edwin Hubble, seen here at the controls of the 100-inch Hooker Telescope at Mount Wilson Observatory, is generally credited with discovering the law that bears his name: that galaxies have red shifts proportional to their distance forms, indicating that we live in an expanding universe. It is only fair to say that his early data was not hugely convincing and that others, such as Lundmark, deserve a share of the credit. (NASA/ESA)

The Astronomers' Stars
A Study in Scarlet

Lynne Marie Stockman

A casual glance at the night sky might lead you to the conclusion that all stars glow white. A little closer study will reveal some subtle colour differences. Reddish Betelgeuse and blue-white Rigel in the constellation of Orion provide a well-known colour contrast. Nearby in the constellation of Auriga, bright Capella glows a warm yellow. Yet sometimes the colours of stars can be quite pronounced and worthy of note.

William Herschel (1738–1822) was born Friedrich Wilhelm Herschel in Hanover in what is now Germany. Like his father, Herschel joined the Hanoverian Foot Guards as a musician and in 1756, the young man found himself with his comrades in Kent, England, augmenting British forces against a possible invasion by the French. During his short time in Kent he took the opportunity to learn the English language. The invasion never materialised and the Hanoverian forces returned to mainland Europe. Soon afterwards, Herschel left the army and returned to England in 1758 where he worked successfully as a musician – copyist, performer, composer – for decades. He became increasingly interested in mathematics and astronomy in the 1770s and soon built his own telescope. With the able assistance of his sister Caroline who joined him in England in 1772, Herschel observed double stars over the course of many years, showing how the components changed position with respect to one another, and conducted a number of deep sky surveys which formed the basis of the later *New General Catalogue of Nebulae and Clusters of Stars* (NGC). He is perhaps most remembered for his discovery of the planet Uranus in 1781 (which won him the Royal Society's prestigious Copley Medal), but among his many other accomplishments were the detection of two moons of Saturn and two of Uranus, and the identification of infrared radiation. From 1807 onwards Herschel was in poor health but agreed to act as the first chairman of the Astronomical Society of London (later the Royal Astronomical Society) two years before his death in 1822 (O'Conner and Robertson 2017).[1]

---

1. References are found at the end of the article.

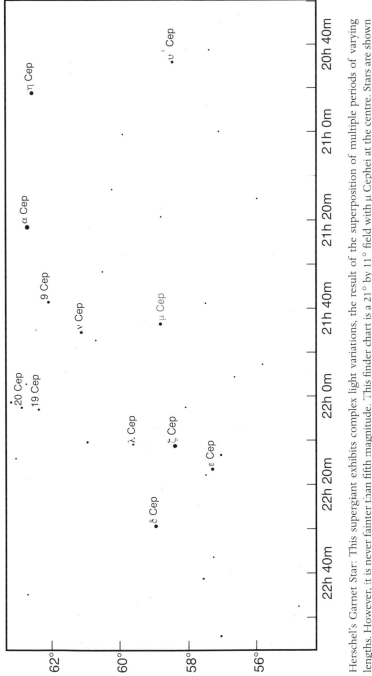

Herschel's Garnet Star: This supergiant exhibits complex light variations, the result of the superposition of multiple periods of varying lengths. However, it is never fainter than fifth magnitude. This finder chart is a 21° by 11° field with μ Cephei at the centre. Stars are shown to magnitude +6.5 with the brightest stars being α Cephei (+2.5), ζ Cephei and η Cephei (both +3.4), and the famous variable star δ Cephei (+3.5–4.4). Coordinates are referred to the epoch J2000. (David Harper)

Herschel's Garnet Star is located in the northern constellation of Cepheus. The name, Latinised as 'Garnet Sidus', appeared in Giuseppe Piazzi's 1803 Palermo star catalogue. It was designated μ Cephei by Johann Bayer in 1603 but John Flamsteed apparently missed it in his star catalogues of 1725 and 1729, and the star's popular name comes from an article written by Herschel in 1783:

> A very confiderable ftar, not marked by FLAMSTEAD, will be found near the head of Cepheus. Its right afcenfion in time, is about 2' 19" preceding FLAMSTEAD's 10th Cephei, and it is about 2° 20' 3" more fouth than the fame ftar. It is of a very fine deep garnet colour, fuch as the periodical ftar o Ceti was formerly, and a moft beautiful object, efpecially if we look for fome time at a white ftar before we turn our telefcope to it, fuch as α Cephei, which is near at hand.

The colour of a star is determined by the temperature on its surface. Stars are classified according to the various characteristics which appear in their spectra and the Harvard classification system – O, B, A, F, G, K, M – is arranged by surface temperature, from hottest to coolest, and thus from blue to white to yellow to orange to red. Each letter class is further subdivided into 0–9 with 0 being the hottest and 9 the coolest. This is accompanied by the Yerkes classification system (later called the Morgan-Keenan or MK classification system) which specifies the luminosity class of a star and is indicated by a Roman numeral. This scheme distinguishes supergiants from giants from dwarfs. There are also some other classes for special types of stars, such as C (carbon stars), D (white dwarfs) and W (Wolf-Rayet stars). Finally, a number of suffixes may be added to indicate any special spectral features, such as 'e' for emission lines and 's' for narrow or sharp absorption lines.

Herschel's Garnet Star is the archetypal M2Ia star, meaning it is cool (M) but at the warmer end of the M class (2). It is also a luminous supergiant (Ia). By comparison, our Sun, spectral type G2V, is a yellow dwarf star, hotter but much smaller than the Garnet Star.

This enormous red star is also variable, exhibiting several periods with reported values ranging from 700 days to over 4,000 days. Astronomer J.R. Hind first noted its changing brightness (Hind 1848):

> The remarkable garnet-stars [sic] in Cepheus appears to be fluctuating in brilliancy. Piazzi calls it – 6, *Taylor* in 1834 – 5–5.6, *Argelander* 1842 Sept. 11 – 3, *Johnson* in Sept. 1845 – 4 or 4.5 while I now make it 3 or at least 3.4. A

very great proportion of telescopic variable stars, which I have noticed in prosecuting our search for planets, are strongly tinged with red.

Its brightness varies between magnitudes +3.2 and +5.4.

The distance to this supergiant star is unknown. Parallax measurements from the *Hipparcos* mission suggest a distance of approximately 1,800 pc whilst results from *Gaia* return a value of around 2,100 pc, but in both instances, the estimated errors are quite large so the numbers are not meaningful. However, it is thought that the star's absolute magnitude is around −7 or −8 which yields a distance less than half that calculated from the reported parallaxes. It is also classified as a runaway star meaning it has an abnormally high space velocity relative to its surroundings.

Stellar models predict that Herschel's Garnet Star weighs in at over 20 solar masses with a diameter more than a thousand times that of the Sun. Additionally, it is surrounded by a shell of ejected gas and dust which includes water vapour. The star is already fusing helium into carbon in its core and is nearing the end of its life. Eventually it will explode as a supernova before collapsing into a black hole.

The only child of William and his wife, Mary Baldwin Pitt, **John Frederick William Herschel** (1792–1871) was also a noted astronomer. From an early age he was taught by his distinguished aunt, Caroline Herschel, and briefly attended Eton College before returning home to be privately tutored. He eventually attended St John's College, University of Cambridge, studying mathematics. He graduated in 1813 and was elected to the Royal Society in the same year. Herschel decided to enter the legal profession upon graduation but soon gave it up, returning to Cambridge as a mathematics tutor. In 1816 he finally joined his ageing father to embark on what would become an illustrious career in astronomy. Like his father, Herschel was involved with the Astronomical Society of London from the beginning, and was elected vice-president of the fledgling group. His first published major astronomical paper continued the work that his father had done on double stars. Herschel arrived with his young family (he had married his cousin Margaret Brodie Stewart in 1829) at the Cape of Good Hope in South Africa in 1834 in order to catalogue interesting astronomical objects not visible from northern latitudes. Whilst there he observed the 1835 apparition of Halley's Comet, discovering that gas was evaporating from it. He returned to England in 1838 and eventually published a book of his observations entitled *Results of Astronomical Observations Made During the Years 1834, 5, 6, 7, 8, at the Cape of Good Hope*. Around this time he also became interested in the new process of photography, writing a number of papers on the infant subject. Herschel was secretary of the Royal Society in 1824 and served as president of the Astronomical Society of London/Royal Astronomical Society three times. He was knighted in 1831 and was made a baronet upon his return from South Africa. He won numerous prestigious awards, including the Royal Society's Copley Medal (1821, 1847) and Royal Medal (1833, 1836, 1840), the Gold Medal of the Astronomical Society of London/Royal Astronomical Society (1826, 1836) and the Prix Lalande (1825). He was survived by all but one of his twelve children, two of whom carried on the Herschel family's astronomical tradition (O'Conner and Robertson 1999).

Herschel observed a number of strikingly red stars over his lifetime. One in particular attracted his attention and he described the sixth-magnitude object as 'A fine ruby star. Pure ruby colour. This is perhaps the finest of my "ruby stars."' (Herschel 1847). *Herschel's Ruby Star* was thought to be variable as early as 1875 (Gore 1875):

> … star must be variable to the extent of at least two magnitudes. It should be watched, as so many of these deep red stars are known variables…As I can find no previous record of any observations on the variability of this star, I propose to name it V *Capricorni*.

It is now known as RT Capricorni. The *General Catalogue of Variable Stars: Version GCVS 5.1* gives a variation in brightness between +6.8 and +8.0 over a period of 423 days. Recent *Gaia* results place the star approximately 420 pc away. Its spectral type is C6,4 which identifies it as a carbon star. (A C6 star is roughly equivalent to a red M-type giant. The '4' signifies the strength of the spectral lines or bands corresponding to certain carbon molecules). Most carbon stars are asymptotic-giant-branch stars which contain more carbon than oxygen in their atmosphere. These stars are past the hydrogen-to-helium fusion process of their youth and are now converting helium to carbon. Convection dredges up carbon from the core and deposits it in the atmosphere. Here it combines with the existing oxygen to form carbon monoxide, leaving the rest of the carbon to preferentially scatter blue light, thus giving the star its distinctive 'sooty' appearance. The Ruby Star is also surrounded by dusty shell of expelled material (Mečina et al 2014).

A different red star caught the attention of another distinguished English astronomer in the nineteenth century. **John Russell Hind** (1823–1895) was born in Nottingham, England, the son of a lace manufacturer. He was apprenticed to a London civil engineer in his teens but his love of astronomy led him to a post at the Royal Observatory in Greenwich in 1840 where he remained for four years. After this he went to work at the private observatory of George Bishop, merchant and gentleman astronomer, in Regent's Park, London. During this time he made numerous discoveries, including such diverse objects as comets, asteroids,

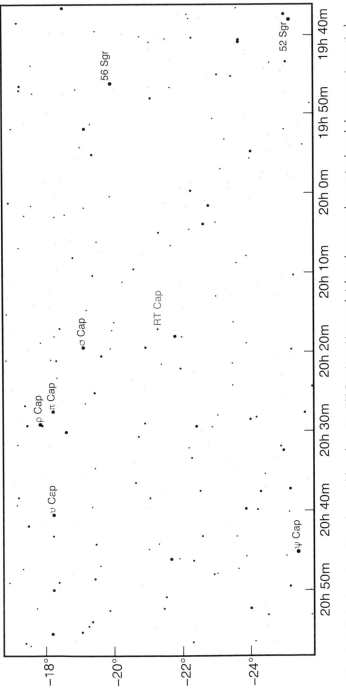

Herschel's Ruby Star: A semi-regular variable carbon star, RT Capricorni is never brighter than seventh magnitude and always requires optical aids to see it. This finder chart is a 21° by 11° field with RT Capricorni at the centre. Stars are shown to magnitude +8.5 with the brightest stars being ψ Capricorni (+4.1), 52 Sagittarii (+4.6), and 56 Sagittarii and ϱ Capricorni (both +4.8). Coordinates are referred to the epoch J2000. (David Harper)

variable stars and nebulae. He was elected to the Royal Astronomical Society in 1844 and to the Royal Society in 1851. Two years later Hind succeeded to the post of superintendent of the Nautical Almanac Office where he remained until his retirement in 1891. Throughout this time he continued to supervise Bishop's observatory. Hind was awarded many accolades and honours during his lifetime, including the Gold Medal from the Royal Astronomical Society, the Royal Medal of the Royal Society, and the Prix Lalande (six times) (Anonymous 1896).

Hind observed a shockingly red star in 1845, noting (Hind 1850a):

Position for 1850 of a Scarlet Star Between Orion & Eridanus.
AR. 4^h 52^m 45^s N. P. D. $105°$ $2'$.

and (Hind 1850b):

I may mention also a remarkable crimson star in Lepus of about the 7th. magn. the most curious coloured object I have seen. The mean place for 1850 is AR. 4^h 52^m 46^s76 $\delta = -12°$ $2'$ $9''3$
[NB: The declination should be $-15°$ rather than $-12°$.]
I found this star in October 1845 and have kept a close watch upon it since.

and (Hind 1850c):

In October, 1845, I found a highly-coloured crimson, or even scarlet, star in *Orion*, far the most deeply coloured object I have yet seen. Its mean place for 1850 is – R.A. 4^h 52^m $46^s\cdot76$.   N.P.D. $105°$ $2'$ $9''\cdot3$.

Hind's Crimson Star is also known by the variable star designation R Leporis. Parallax values from *Gaia* give it a distance of approximately 420 pc from Earth. The spectral type of Hind's Crimson Star is C7,6e which characterises it as a carbon star with emission lines in its spectrum. The Crimson Star is a Mira-type variable. Named for the prototype star in Cetus, Mira variables are very cool and thus very red, with pulsation periods of over 100 days and large-amplitude brightness changes of two magnitudes or more. They have distended atmospheres (200–300 times the radius of the Sun) and high luminosities (3,000–4,000 times the luminosity of the Sun). They are also undergoing rapid mass loss, thus enriching the nearby interstellar medium with heavy elements. Many of them are surrounded by dusty shells and some form planetary nebulae before they evolve into white dwarfs. According to the *General Catalogue of Variable Stars: Version GCVS 5.1*, the pulsation period of

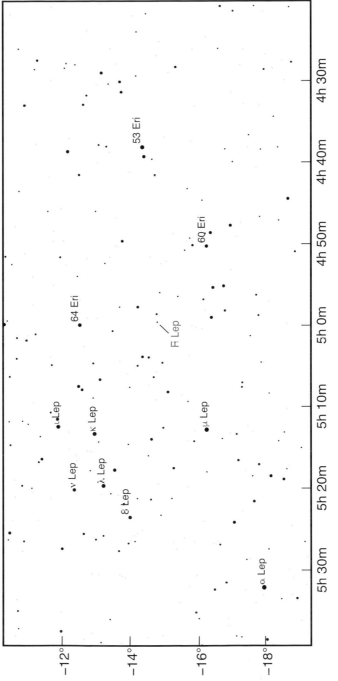

Hind's Crimson Star: This carbon star is a Mira-type variable, sixth magnitude at its brightest and twelfth magnitude at its faintest and reddest. Binoculars or a telescope are normally required to observe this object. This finder chart is a 21° by 11° field with R Leporis at the centre. Stars are shown to magnitude +8.5 with the brightest stars being α Leporis (+2.6), μ Leporis (+3.3) and 53 Eridani (+3.9). Coordinates are referred to the epoch J2000. (David Harper)

Hind's Crimson Star is 427.07 days, with a variation in apparent brightness between magnitudes +5.5 and +11.7.

Like all red variable stars, Hind's Crimson Star appears reddest in appearance when it is at its dimmest. With its long period and large amplitude changes, it is an excellent target for the amateur variable star observer and photographer.

Red stars have captured the imagination since the advent of the telescopic age. As Irish astronomer John Birmingham said in 1879:

> The Red Stars must be considered as a class of heavenly bodies particularly worthy of attention; for not alone, as compared with the other stars, do they seem to differ most widely in constitution from our own sun, but they show a peculiar inclination to periodic change, while some of the most noted Variables are found amongst them.

Further Reading

For more information on these and other historical red stars, see 'William Herschel and the "Garnet" Stars: μ Cephei and More' by Dr Wolfgang Steinicke and published in 2015 in the *Journal of Astronomical History and Heritage*, volume 18, number 2, pages 199–217. It is also available online at the author's web site **klima-luft.de/steinicke**

Acknowledgements

This research has made use of NASA's *Astrophysics Data System Bibliographic Services*, operated at the Harvard-Smithsonian Center for Astrophysics, Cambridge, Massachusetts, USA, and the SIMBAD astronomical database, operated at CDS, University of Strasbourg, France. The author would like to thank Dr David Harper for his enthusiastic encouragement and helpful comments.

References

[Anonymous] 1896, 'Obituary: John Russell Hind', Monthly Notices of the Royal Astronomical Society, 56, 200–205.

Birmingham, J. 1879, 'The Red Stars: Observations and Catalogue', The Transactions of the Royal Irish Academy, 26, 249.

Gore, J.E. 1875, 'Note on a new Variable Star in Capricornus', Monthly Notices of the Royal Astronomical Society, 36, 70–71.

Herschel, John F.W. 1847, Results of Astronomical Observations Made During the Years 1834, 5, 6, 7, 8, at the Cape of Good Hope; Being the Completion of a Telescopic Survey of the Whole Surface of the Visible Heavens. Commenced in 1825., (London: Smith, Elder and Co.), 449.

Hind's Crimson Star is caught by the camera on 6 September 2015 at tenth magnitude. This image is approximately 25 arc-minutes wide and shows objects to at least magnitude +17. The two 'bright' blue stars flanking the Crimson Star are both A-type; ninth-magnitude dwarf HD 31901 is on the right and tenth-magnitude subgiant HD 32119 is on the left. Above and to the right of the Crimson Star is the yellowish tenth-magnitude star BD−14 1009. All of the other stars in the photograph are eleventh magnitude or fainter. (Damian Peach)

Herschel, William. 1783, 'On the proper Motion of the Sun and Solar System; with an Account of several Changes that have happened among the fixed Stars since the time of Mr. Flamstead: III Stars newly come to be visible', *Philosophical Transactions of the Royal Society of London*, 73, 257.

Hind, J.R. 1848, 'Schreiben des Herrn Hind an den Herausgeber', *Astronomische Nachrichten*, 27, 373–374.

Hind, J.R. 1850a, 'Auszug aus einem Briefe des Herrn Hind', *Astronomische Nachrichten*, 30, 257–258.

Hind, J.R. 1850b, 'Auszug aus einem Schreiben des Herrn Hind an den Herausgeber', *Astronomische Nachrichten*, 30, 275–276.

Hind, J.R. 1850c, 'Notes on the British Association Catalogue', *Monthly Notices of the Royal Astronomical Society*, 10 (6), 141.

Mečina, M., Kerschbaum, F., Groenewegen, M.A.T., and 8 others. 2014, 'Dusty shells surrounding the carbon variables S Scuti and RT Capricorni', *Astronomy & Astrophysics*, 566, A69.

O'Connor, J.J., Robertson, E.F. 1999, 'John Frederick William Herschel', *MacTutor History of Mathematics*, **mathshistory.st-andrews.ac.uk/Biographies/Herschel** (accessed 20 January 2021).

O'Connor, J.J., Robertson, E.F. 2017, 'Frederick William Herschel', *MacTutor History of Mathematics*, **mathshistory.st-andrews.ac.uk/Biographies/Herschel_William** (accessed 20 January 2021).

Frank Drake and His Equation

David M. Harland

This article looks back at a pioneering experiment in the early 1960s to 'listen' for radio signals from civilisations around other stars. No one knew what to expect...

-oOo-

Frank Donald Drake was born on 28 May 1930 in Chicago, Illinois, to Richard Carvel Drake and Winifred Pearl Thompson, who met while at the University of Illinois. His father went on to work for the city as a chemical engineer. There were three children, Frank being the eldest, and the family abided by the morality of the Hyde Park Baptist Church, which was on the campus of the University of Chicago.

Drake's science education at school was bolstered by frequent visits to the Museum of Science and Industry in Chicago. It introduced him to all manner of wonders, particularly astronomy, which he followed up with visits to the Adler Planetarium. The realisation that the Sun is just an average star in a galaxy of billions made him wonder whether there were alien civilisations, and what they might be like.

On graduating from high school, Drake's ambition was to become an aerospace engineer. Armed with a scholarship from the Naval Reserve Officer Training Corps, he decided to go Ivy League, accepting an offer from Cornell University in Ithaca, New York, "because it had girls". But he soon realised that aeronautical engineering wasn't for him. He was more interested in electronics. After scoring a perfect 100 in the basic electronics course, the first student ever to do so, he switched his major to electronics.

Still fascinated by the possibility of extraterrestrial life, in his second year he took an introductory astronomy course and was smitten by the sight of Jupiter and its moons through a telescope.

In 1951, late in Drake's third year, he attended a lecture in which the renowned Chicago astrophysicist Otto Struve explained that the profile of the dark absorption lines in starlight implied that massive stars with spectral types O, B, A and F rotated rapidly whereas smaller dwarfs of types G, K and M rotated slowly. Struve interpreted the slower rate of rotation as indicating that a star had transferred angular momentum to a system of planets.[1] This was an eye-opener for Drake, since it implied that planetary systems were common for stars similar to the Sun and therefore possible abodes for alien civilisations.

Drake graduated in 1952 with honours in engineering physics and a desire to become an astronomer, but first he had to spend three years in the Navy to repay his scholarship. After training at electronics school, he was put in charge of electronics maintenance on board a warship. In 1955 Drake was accepted by the Harvard astronomy graduate program. In view of his electronics experience, the university gave him a summer job working on a radio astronomy project, an accident of circumstances that would seal his fate.

Radio astronomy was a new discipline. The first detection of celestial radio waves by Karl Jansky at Bell Telephone Laboratories in Holmdel, New Jersey, in 1931, was made inadvertently while he was seeking the source of 'static' which impaired the quality of radio transmissions. He identified the source as the Milky Way, and then tackled other work for his employer. Nothing further was done until radio enthusiast Grote Reber built a large antenna at his home in Wheaton, Illinois, and mapped the strength of the emission across the Milky Way, publishing his results in 1944 with the assistance of Otto Struve.

Leaving Harvard with a Ph.D. for radio observations of the open star cluster known as the Pleiades, Drake joined the nascent National Radio Astronomy Observatory (NRAO), Green Bank, West Virginia, in April 1958 as a staff astronomer. It lies in a wooded valley of the Allegheny Mountains, protected from much of the radio interference from nearby cities. The plan was to construct a fully steerable dish antenna having a diameter of 140 feet, which at that time was far larger than anyone had elsewhere, but because that would take several years (and in fact it was not finished until 1964) it was decided to build a simpler 85-foot dish in the interim to test instruments and enable an observational program to get underway as soon as possible.

Frank Drake in the early 1960s. (NRAO/AUI/NSF)

The 85-foot antenna at the National Radio Astronomy Observatory at Green Bank in West Virginia. (NRAO/AUI/NSF)

On the 85-footer's completion the following year, Drake realised that if aliens on planets orbiting nearby stars had a broadcasting capability similar to ours, then the antenna, if it were to be equipped with one of the new solid-state detectors that promised a factor of 100 improvement over earlier models, would be capable of detecting their signals.

At that time, Lloyd V. Berkner, the head of Associated Universities, Inc., the NRAO's parent organisation, was running the observatory on an interim basis. On being shown Drake's calculations, Berkner told the young astronomer to give it a go. This was confirmed by Otto Struve when he became director shortly thereafter.

Drake chose the name Project Ozma in homage to the princess in the Lyman Frank Baum children's story *Ozma of Oz*, the first part of which was *The Wonderful Wizard of Oz*, because it envisaged "a land far away, difficult to reach, and populated by strange and exotic beings".

Philip Morrison. (Cornell University)

To Drake's astonishment, in September 1959 a paper appeared in the journal *Nature* entitled 'Searching for Interstellar Communications'.[2] In their independent analysis, Giuseppe Cocconi and Philip Morrison of Cornell had, like Drake, realised that the newest radio telescopes were sufficiently sensitive to detect signals like those we were broadcasting into space. In calling for a search, they said, "The probability of success is difficult to estimate, but if we never search, the chance of success is zero."

In a lecture at the Massachusetts Institute of Technology a few weeks later, Struve announced that the NRAO was already preparing such a search.

As regards the wavelength, Drake and (independently) Cocconi and Morrison, concluded that the search should be at wavelengths close to that at which interstellar neutral hydrogen atoms radiate as their single electron inverts its direction of axial spin (namely 21 cm, a frequency of 1,420 megahertz) on the presumption this would be a logical choice for anyone attempting interstellar radio communication.

While the telescope was being assembled, Drake designed a single-channel receiver with a bandwidth of 100 hertz which would scan a 400 kilohertz band centred on the marker frequency. It needed a very stable state-of-the-art oscillator in order to scan that narrow frequency range. An electrical engineer from England, Ross Meadows, was hired to help with assembly and testing.

One day, Dana W. Atchley Jr., president of Microwave Associates, Inc., near Boston, who had heard of the project from Struve's talk, phoned Drake and offered a very capable new signal amplifier that was not yet on the market; Drake eagerly accepted.

As targets for the experiment, Drake selected two nearby solar-type stars. Epsilon Eridani lies in the constellation of Eridanus (the River). Its K2V spectrum meant it was significantly cooler than the Sun, with one-third the luminosity. Its parallax on the sky put it 10.5 light years away. The profile of the absorption lines in its spectrum indicated a rotation period of 11 days, which (by Struve's logic) implied the presence of planets. Tau Ceti is in the constellation of Cetus (the Whale). Its G8V spectrum meant it was only slightly cooler than the Sun, with about half the luminosity. It was slightly farther away, at 12 light years. Its 34-day rotation period was a strong indication that it possessed a planetary system.

The two stars were located about two hours apart in terms of Right Ascension, so an observing run would begin with Tau Ceti, then when that was low in the sky the antenna would be slewed eastward to pick up Epsilon Eridani.

---

2. 'Searching for Interstellar Communications', Giuseppe Cocconi and Philip Morrison, *Nature*, vol. 164, pp. 844-846, 19 September 1959.

Observing began on 8 April 1960, with Drake assisted in the control room by Kochu Menon, a recruit from the Harvard radio astronomy graduate program, and a pair of volunteers in the shape of the NRAO's two female students, Ellen Gundermann and Margaret Hurley. In those primitive times the primary output was a squiggly line on a paper chart recorder, but the signal was also stored on magnetic tape and played on a loudspeaker.

With only the occasional 'false alarm' from terrestrial interference to raise momentary excitement, the task soon became rather tedious. From time to time, visitors would pop in to see how things were going, and it was reported in *Time* magazine and *Saturday Review*.

After a month there was a pause to allow the antenna to be used for other work, then it finished with a second month. In all, 200 hours' of data had consumed thousands of meters of chart paper and magnetic tape without finding any trace of an intelligent signal of extraterrestrial origin.

Drake wasn't disheartened by a 'negative' result. After all, it would have been truly remarkable if the first 'listening' exercise had found an unambiguous signal from an alien civilisation. He was conscious of the fact that if a signal was being broadcast by either of the targets, it could easily have been below the threshold of his instrument, despite the size of the dish and the use of a state-of-the-art detector and amplifier. And of course, the negative result might have arisen because the aliens were transmitting on some other frequency or were taking a vacation. Ozma provided operational lessons on how to conduct a future search, most notably ways to eliminate signals of terrestrial origin

A year later, in the summer of 1961, Drake received a call from J. P. T. Pearman, a staff member of the National Academy of Sciences' Space Science Board, who invited him to organise a gathering "to examine, in the light of present knowledge, the prospects for the existence of other societies in the galaxy with whom communications might be possible; to attempt an estimate of their number; to consider some of the technical problems involved in the establishment of communication; and to examine ways in which our understanding of the problem might be improved".

Struve promptly agreed that Green Bank should host this workshop on 'Intelligent Extraterrestrial Life', and said he would chair it. Most of the people invited to participate signed up immediately.

In addition to Struve, Drake and Pearman, the attendees were Philip Morrison of Cornell, who had co-authored the *Nature* paper; Dana W. Atchley Jr., who provided the amplifier for Ozma; Bernard M. Oliver, a senior figure at the Hewlett-Packard Corp., who had shown an interest in Ozma; Melvin E. Calvin, a biochemist at the University of California, Berkeley, who had established that sunlight acts on the chlorophyll in a plant to fuel the manufacturing of organic compounds, rather than on carbon dioxide as had been believed; Carl Sagan, an astronomer with a keen interest in biology who had recently gained his doctorate from the University of Chicago, was on the Space Science Board's Committee on Exobiology (established in 1958), and was now working with Calvin on experiments to investigate how life might have begun on Earth; John C. Lilly, a neuroscientist who was at that time studying consciousness and communication, in particular with dolphins at Coconut Grove, Florida; and Su-Shu Huang, one of Struve's former students, now working for NASA, who had studied the 'habitable zones' associated with different types of star.

To introduce the theme of the workshop, Drake asked himself what kind of information would be required to investigate the prospects for extraterrestrial life. After making a list of relevant points, he realised they were interdependent and he could express them as a probabilistic equation for N, the number of communicative civilisations currently in the galaxy:

$$N = R \cdot f_p \cdot n_e \cdot f_l \cdot f_i \cdot f_c \cdot L$$

where

R = the average rate at which suitable stars are formed
f_p = the fraction of those stars that possess planets
n_e = the average number of planets per star which are conducive to the development of life
f_l = the fraction of those planets on which life develops
f_i = the fraction of life-bearing planets that develop intelligence
f_c = the fraction of intelligent species that broadcast radio
L = the length of time that such a civilisation remains detectable.

This provided an excellent means of structuring the debate, enabling the participants to estimate values for the various terms. The workshop got underway on Wednesday, 1 November, 1961.

First, they agreed that they were interested primarily in "good suns", meaning those that were neither too hot nor too cool, but similar to our Sun. The astronomers suggested that the R term was simply a matter of dividing the number of solar-type stars in the galaxy (estimated at 10 billion) by the average lifetime of such a star (estimated at 10 billion years), giving a value of 1. Although this was only an approximation, what was important was that R was neither enormous nor minuscule.

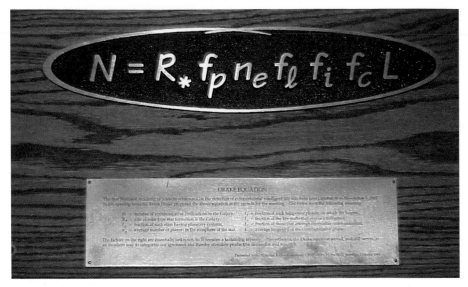

Although the blackboard on which the famous equation was written is no longer in the room at the NRAO where the workshop was held, it is celebrated by this plaque. (SETI League)

Next was the f_p term. Roughly half of dwarf stars with spectral types F, G and K were members of binary systems, and so were not relevant. Not every solitary star would be attended by planets, because Struve's study of the rates at which stars rotated implied that 'fast rotators' were unlikely to have planetary systems. It was therefore decided that f_p was in the range 0.2 to 0.5.

For n_e, it was noted that in the case of our solar system the 'habitable zone' had Venus at its inner edge, Earth in the centre, and Mars at the outer edge. At that time, no stars had been firmly established to possess planets, and because the only system we knew of was our own, it had to be regarded as typical, so it was decided to be generous and put n_e in the range 1 to 5.

In marked contrast to the astronomical questions, where there was at least a reasonable basis for inference, the terms relating to biology and intelligence were far more contentious.

For how many of the planets orbiting within the habitable zone of a star would develop life, the experiments by Calvin and Sagan appeared encouraging, in that they suggested that life was the natural outcome of chemical reactions and was likely to be commonplace, so f_l was set at 1.

The second day was interrupted to celebrate the news that Calvin had been awarded the 1961 Nobel Prize for Chemistry for his work on chlorophyll.

The discussion of how many life-bearing planets would give rise to intelligence covered a lot of ground, decided this was highly likely, and set f_i at 1.

Opinions regarding how many intelligent species would be capable of broadcasting signals differed greatly. Dolphins were intelligent but had not developed radio. It was decided to set f_c in the range 0.1 to 0.2.

At this point it was noticed that all of these factors tended to cancel out, yielding a value of 1. The number of communicative civilisations in the galaxy would therefore be determined entirely by the final term, L, which was the longevity of such a civilisation. This depended on how optimistic,

Enrico Fermi in the mid-1940s. (US Department of Energy)

or indeed pessimistic, a view was taken of the likelihood of a newly intelligent species annihilating itself. Perhaps it might last only a thousand years, but it might survive for millions of years. By this logic, there might be thousands to millions of civilisations broadcasting radio signals right now.[3]

All that could be said for sure was that the value of N was at least 1, namely ourselves, armed with radio telescopes. The challenge for what became known as the Search for Extraterrestrial Intelligence (SETI) is to find at least one interstellar neighbour because, as Morrison put it, that would "transform the origin of life from a miracle to a statistic".

Although the terms selected for the equation and their likely values were all open to further debate, as a rationale for that first discussion it served its purpose.

The 'Drake Equation' gained an iconic status, but as the man himself has observed, it didn't represent any great intellectual insight. In fact, it is a narrow example of a much broader analysis.

At the Los Alamos National Laboratory in New Mexico in 1950, the renowned physicist Enrico Fermi mused on the apparent contradiction between the high

---

3. It was later argued that L should be not the lifetime of the civilisation but the time it is detectable as a radio source. As the efficiency of communications systems improves, thereby reducing signal 'leakage', they will fade below our detection threshold.

probability of there being extraterrestrial civilisations and the lack of evidence for them. This has become known as the Fermi Paradox.

The universe is now known to be almost 14 billion years old but the Sun was formed by the collapse of an interstellar cloud a mere 5 billion years ago, so many of the solar-type stars in the galaxy are much older than the Sun. If many of these stars possess planets on which intelligent life developed and created a means of travelling between the stars, then even advancing at sub-light speed such a civilisation could explore the entire galaxy in a few million years, in which case where are they? The Drake Equation does not include the possibility of aliens colonising the galaxy.

All the possible solutions are enticing. Perhaps intelligent life is rare. Indeed, perhaps, despite the youth of the Sun, we are the first. Perhaps the majority of civilisations remain within their own planetary systems. Perhaps independent civilisations communicate with one another. If so, we might one day detect them, with Drake's experiment being only the first attempt.

Subsequent SETI expanded the range of wavelengths studied to span the 'waterhole', delimited at one side by the 21-cm neutral hydrogen line and at the other side by the 18-cm hydroxyl-radical (OH⁻) line. The moniker was proposed by Bernard Oliver in 1971 because these are the dissociation products of water, which is considered crucial to life. But it has also been proposed that aliens would be more

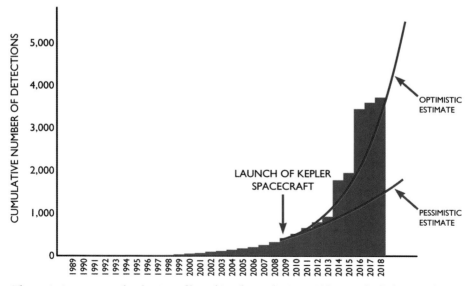

The optimism expressed at the time of launching the Kepler Space Telescope for finding exoplanets proved to be fully justified. (Data from NASA and The Extrasolar Planets Encyclopaedia of exoplanet.eu)

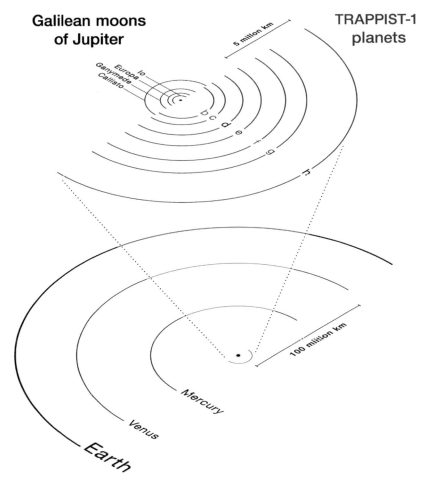

The ultra-cool red dwarf star designated TRAPPIST-1 is located 40 light years away in the constellation Aquarius. The orbits of its planets are so compact that they are comparable with the satellites of the Jovian system. With so many planets in such close proximity, it was possible to accurately measure their masses. Five are similar in size to Earth and two are intermediate between Mars and Earth. The star has 8% the mass of the Sun, so all of the planets are 'temperate' and three are within the 'habitable zone'. (ESO/O. Furtak)

likely to utilise laser beams for increased bandwidth. If so, we would not be able to detect a signal unless it was aimed at us or we had the remarkable good fortune to lie on the line of sight between two communicating civilisations.

Many years elapsed before we developed the means of detecting the presence of planets around other stars. The rate at which they are being detected is accelerating,

The Allen Telescope Array at the Hat Creek Radio Observatory north of San Francisco. (SETI Institute)

with several thousand now known. It came as a considerable surprise to discover that our planetary system is far from being typical.

As our investigations determine which stars possess planets in their habitable zones, this can be used to target future SETI. For example, an ultra-cool red dwarf with 8% the mass and 11% the radius of the Sun, 40 light years away in the constellation Aquarius would not have been considered a "good sun" by Drake, but we now know it has a system of seven planets so compact as to be similar in size to the Jovian satellites. All of these planets are small rocky bodies, three of them located in the habitable zone. In 2016, the Allen Telescope Array scanned this system using 10 billion radio channels seeking an alien presence, with negative results.

The search continues...

Further Reading

Is Anyone Out There? The Scientific Search for Extraterrestrial Intelligence by Frank Drake and Dava Sobel, Delacorte Press (N.Y.), 1992.

If the Universe Is Teeming with Aliens ... WHERE IS EVERYBODY?: Seventy-Five Solutions to the Fermi Paradox and the Problem of Extraterrestrial Life by Stephen Webb, (2nd ed.), Springer, 2015.

Other Worlds. The Search for Life in the Universe by Michael D. Lemonick, Simon & Schuster, 1998.

Remote Observing and Imaging

Damian Peach

Introduction

The age of the internet has brought about many changes over the last twenty years or so, and Astronomy has been no exception. The concept many have of an amateur astronomer is someone outside in their garden, with telescope pointed skyward in eager anticipation of the wonders that await them through the eyepiece.

While only twenty years ago this was almost exclusively true, modern technology (specifically the age of high speed internet access) has brought amateurs closer than ever before to their professional counterparts. Even owning a telescope of your own is no longer a necessity to do worthwhile astronomy and astrophotography. Logging onto remote telescopes via the internet has become an increasingly popular way for both novice and serious astronomers both amateur and professional to conduct observations.

How Remote Observing Works

Most remote observatories such as Chilescope or iTelescope work on a points per telescope per hour system. This means you purchase points and any given telescope costs a certain amount of points per hour to use. Various membership plans are available, all of which are a monthly recurring subscription, anything from as little as £12 per month to as high as £600 per month. Telescope points per hour usage rates vary depending upon your membership plan and whether the moon is present in the sky at the time of your observation. Those paying more per month benefit from lower points per hour usage rates on all telescopes.

Completely free trials of some services such as Itelescope are available whereby users can take control of smaller telescopes and take an image. This is a great way to see how the service operates without making any commitment. Membership plans can usually be cancelled or changed at any time. Below you will find an example breakdown of two different iTelescope plans and how the expense varies:

Membership Plan	Telescope	Cost
Plan 40 (40 points or £26 per month)	Telescope 9 (12.5 inch RCOS with STL-11k)	132 points per hour
Plan 290 (290 points or £200 per month)	Telescope 9 (12.5 inch RCOS with STL-11k)	74 points per hour

The iTelescope observatory at Siding Spring, Australia. This remote observatory along with several others is among the most popular world-wide for the amateur observing community. (Damian Peach)

The above gives examples of lower and a higher end plans and how the expense varies on a mid-range telescope on a moonless night. Discount varies depending upon the lunar phase, with up to 50% cheaper rates during times when the moon illumination is greater than 75%. It is also clear from the above how those paying more each month obtain much cheaper rates, thereby allowing more time to make observations. You only pay for the actual time the CCD camera is active, so all time taken up during slewing, focusing and so on is not charged, which means you are only using points for actual imaging time and nothing else.

Observatories and Locations
The primary remote observatories available to amateurs are those hosted by Slooh, iTelescope and Chilescope. Others such as Telescopelive have also recently become active. All work on much the same premise of renting time to observe with various types of membership available.

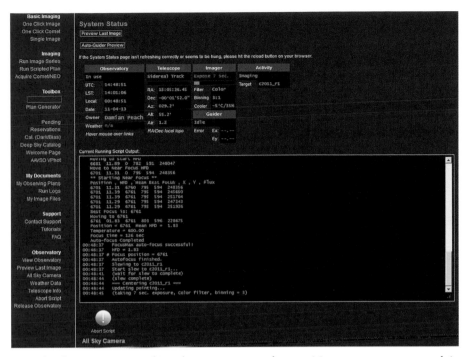

Example of an imaging run taking place on a remote telescope. Many systems you can watch in real time and watch data come off the telescope as it is downloaded. (Damian Peach)

All remote observatories are located at dark sky location in various countries from Chile to Australia. Both northern and southern hemispheres are well covered. iTelescope has sites in both hemispheres in Australia and the USA.

Telescopes on these networks vary greatly in size and each is geared toward different tasks. Small fast refractors such as the popular Takahashi FSQ106 are used in line with large format high quality CCD cameras to offer telescopes suited to capturing images of several degrees of the sky at any one time. On the other hand you also have larger aperture 17″ and 20″ Planewave CDK telescopes equipped with high end SBIG or Finger Lakes CCD cameras, these being geared toward shooting faint galaxies and comets or doing variable star photometry. Some observatories such as Chilescope offer using very large telescopes up to 1m aperture, and additionally also offer remote planetary imaging – the only observatory worldwide to do so at the time of writing.

The Chilescope observatory high in the Atacama Desert. This facility offers a great range of top level equipment, including remote planetary imaging – the only remote observatory to offer this service. (Damian Peach)

In-between these extremes there are telescopes such as 250mm F3.6 astrographs, and 150mm apochromatic refractors. Basically a telescope to suit almost any task you can think of.

All of the telescopes on these networks are mounted on very high quality mounts. These offer extremely accurate tracking and pointing. As mentioned all of the CCD cameras are high end models from either SBIG or FLI and all deliver high quality images. All telescopes are equipped with a wide range of filters from typical RGB to photometric filter sets. A couple of the smaller telescopes are equipped with one shot colour cameras for easy colour astrophotography and these deliver surprisingly good results and are also cheap to use.

Full detailed web pages are provided on each telescope including its field of view so you know all of the required details to best chose a telescope for your purpose.

Regardless of the telescope data captured during any run is transferred to your own private FTP account in high quality FITS format and you retain full copyright on the images, so they are yours to use as you wish. Calibration dark frames and flat field frames are also provided as are fully calibrated images that are automatically dark and flat corrected before transfer to your account.

Booking Time and Taking Images

Most remote services offer a straightforward booking system where you can reserve time slots of your choice. Some telescopes can often become quite busy, especially around New Moon periods so booking time slots well in advance is advisable! It is also worth perhaps booking a bit more time than you need to ensure your reservation completes.

Most remote systems allow booking automatic imaging plans that you have constructed in advance usually using the online web interface provided by the service.

Once you have decided on your target and set your exposure times, filters and other settings you are ready to start capturing. Before capturing begins most remote telescope will point to a focus star and attempt to autofocus. This typically takes around two to three minutes. Sometimes (and this is pretty rare) the process will fail (for a variety of reasons) and the telescope will default to using pre-defined filter offsets. These often work, though of course you can abort the run at any time if you are not happy and want to try again.

Remote telescopes also plate solve all exposures. This means it matches the star field the CCD is imaging to its database to ensure it's pointed exactly where you

Mars in conjunction with the Lagoon and Trifid Nebulae obtained via a remote session at iTelescope Siding Spring. Using remote telescopes allows you to capture unusual events like this that you may otherwise miss completely. (Damian Peach)

tell it. Pointing is extremely accurate and if you have good star charting software (such as Guide 9) you can choose exactly where to point the telescope to perhaps best frame a group of objects.

Once the telescope has focused and slewed to your object it will begin capturing data. Typical most users will shoot several sub exposures, so perhaps 10 × 5 minute exposures. You'll see exactly what the telescope is doing via the live script output and as mentioned above the live preview images are usually available to check all is going ok.

Once a remote imaging run completes the system will automatically log you off and your account will be debited the amount of points used during capture. Your data will then be transferred to your FTP account. You can then log onto your FTP account via an FTP client to download the data. All data is provided in FITS format so compatible with almost any software used for processing astronomical CCD images.

A Powerful Resource

Having been an observer and imager of some 20 years now and having used and owned many telescopes and cameras and observed at many different locations around the world I would certainly consider Myself a very practical astronomer, and initially I really didn't see much appeal in using remote telescopes to make observations. It felt a bit "detached" from really being an astronomer out there under the night sky.

However, over the last several months my views have changed greatly and I have come to appreciate what tremendously valuable and powerful resources remote observatories are. No longer are truly dark skies reserved for those lucky enough to live at favourable locations. Through the use of remote telescopes the quality of these skies are brought to you. Light pollution is a major problem for the vast majority of observers and many are unable to easily escape it. Remote observing services enable you to use quality telescopes located at first class observing sites likely far beyond anything most could find at home.

But of course this comes at price, and for most not an especially cheap one. There is no escaping that remote observing can be expensive. But so can purchasing a similar set-up of your own. Some of the systems on the network would cost well over USD $60,000 to purchase privately – a massive sum to most people, and many private telescopes purchased by most don't really get used that much, either through poor weather or busy family lives. You start to question is spending many thousands on a system of your own really the right course of action? A subscription to a remote observatory service offers an almost completely hassle and stress free experience, and you have the massive bonus of the telescopes being situated at sites far beyond the sky quality most will find at or near home. It is certainly food for thought!

Where something as complex as astronomical imaging is concerned, things always go wrong at one time or another. Most remote imaging providers offer points refunds if for any reason you are not happy with the data captured or something has gone wrong during capture. I have personally had this happen many times and issues or refunds have always been swiftly resolved. All the remote services I have tried provide excellent customer service, and enquirers are always responded to.

Some remote observatories also offer the option of hosting your own remote telescope. Cost however is significant, and teams of observers often join together to run a remote observing system as a team. There are many successful examples of this, especially Deep Sky imaging.

Mars on 17 October 2020 obtained during a remote observing session using the Chilescope 1m F/8 telescope. Very fine details can be seen on the surface including the huge Olympus Mons volcano at right. (Damian Peach)

Final Thoughts

Remote astronomy has really re-kindled my own interests in deep sky and comet imaging. Skies here at home are often cloudy and light polluted and remote imaging represents an effective way of making high quality observations beyond what could be achieved at home. For me it also represents a more effective use of spending money on astronomy. I enjoy actually making observations rather than admiring a telescope so rather than spending a vast sum on a system located under a mediocre quality sky a subscription to a remote observing service enables very high quality observations to be made so money is spent in actually doing astronomy rather than on equipment. Everyone will of course feel differently on this matter but this is certainly a point to think about carefully. For many it will most likely act as a supplement to observations made with their own telescopes, and in this respect remote astronomy works superbly well.

Most remote imaging providers offer a mostly hassle free experience. Any niggles are minor. Some telescopes can become heavily booked, and I think it could be a

Comet SWAN on 2 May 2020 captured via remote session using Chilescope's super wide field 200mm F/2 lens system. (Damian Peach)

good idea to set a user time limit per night per telescope at busy times to enable everyone to make use of the larger telescopes not just a select few booking blocks of several hours at a time. This of course will become less of a problem as more telescopes come online easing the demand.

It will also be great to see a few telescopes dedicated to certain tasks such as planetary observing, spectroscopy or evening/morning comets (which can often be located very low in the sky) often beyond the minimum elevation limit of many of the telescopes at present. As the network expands I am sure it will become even more useful and valuable and continue to represent a tremendously powerful and valuable resource to the entire astronomical community.

Remote Observatory Websites
iTelescope: **www.itelescope.net**
Chilescope: **www.chilescope.com**
TelescopeLive: **telescope.live**
Slooh: **www.slooh.com**

Skies over Ancient America
The Dawn of Sky Watching for Pre-Columbian Civilizations

P. Clay Sherrod

No scientist or anthropologist can establish the date when prehistoric people first looked skyward in their attempts to reason the existence of all that they saw in the heavens, so spectacularly dense with stars and celestial wonders enhanced by the inky blackness of the non-illuminated night time sky.

There is no doubt that the sky held mythical stories, mysteries and certainly burning questions as to the objects that were seen, even to the first humans emerging in isolated locations around the Earth some three million years ago. Remains of our first ancestors who had developed skills for tool-making, and hence the capacity for creative and logical thought have been located in Morocco, dating to possibly 350,000 years before present (BP).[1]

It was far later that any sign of Homo sapiens appeared in the western hemisphere and Americas; in fact it would be nearly a quarter million years before evidence of migration into our western hemisphere took place by the early *Clovis* nomads, this procession beginning near the end of the Laurentide Ice Age regression.[2] These scattered explorers passed into the Americas from Asia via the Bering Strait and reached destinations in a string-like pattern from what is now

A single bone artefact from the Dar es-Soltan 1 Cave in Morocco is the oldest known well-dated specialized bone tool (knife) associated with the Aterian culture from the Middle Stone Age. (Natural History Museum of London, by Silvia Bello and Mohammed Komal, [Fotokam, Morocco])

1. Yong, Ed., *The Atlantic*, June 2017: **www.theatlantic.com/science/archive/2017/06/the-oldest-known-human-fossils-have-been-found-in-an-unusual-place/529452**

2. Pringle, Heather, 'The First Americans', *Scientific American*, November 2012: **www.scientificamerican.com/article/the-first-americans**

northern Alaska, southward through the western United States travelling as far as Chile where their crudely fashioned stone and bone implements as well as human faeces reveal their presence 14,800 years BP.

Rock Art: Dawn of Ancient Astronomy in the Americas

The Paleolithic period consumed the activities and concentration of early mankind; ten thousand years ago, "making a living" meant hunting and gathering food and resources for survival. There were no time clocks, no commutes, no free time for developing concepts of innovation. You wandered and hunted or you did not make a living – you did not live. It was not until the advent of agriculture in the western hemisphere when early humans were afforded TIME, and time is essential for creativity and invention.

These earliest humans in the Americas lived neither in "tribes" nor in shelters; they lived in rock shelters which afforded some protection and warmth, yet did not confine them in the event escape was needed as would a deep and dark cave. It was not until the advent of agriculture that humans could "stay home and grow their food" rather than chasing it, and only then were community cultures established.

In spite of the overwhelming and many times unsuccessful hunting excursions of the nomads, there was likely campfire time: a brief period in which stories were told and legends began. Just as we might do today in our books and digital writings, the use of *illustrations* was key to getting the point across to anyone who might listen. Such drawings and carvings became prevalent throughout the human expanse of the globe, emerging as early as 20,000 BC in the Lascaux Caves of France and found worldwide as it continued until modern times.

Perhaps quite accidentally, a long period of "rock art" emerged for nearly ten thousand years when a cave dweller may have moved a disintegrating stone away from a fire ring, only to find sticky reddish pigment left from the iron of the stone and the grease of dripping cooking meat. As we might do today, perhaps this person merely wiped his or her hand across the stone face of the rock shelter, leaving behind an

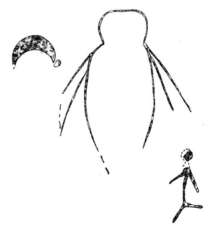

A prehistoric pictograph from hidden rock shelters of Arkansas which may depict the great supernova of 1054 AD. (P. Clay Sherrod, *Motifs of Ancient Man*)

image of a HAND – the impact of which could have been alarmingly shocking as would seeing the first reflection of yourself in a mirror. From that point we might imagine the rapid development of "cave art" or prehistoric rock art as it is know to archaeologists. It was, in essence, the first 'selfie.'

Quickly the human mind was able to convert thought into motor skills and render an image of the hunt, perhaps a wild animal and the danger of its pursuit; an image of the setting sun might indicate that this rock shelter was warmed by the southern sun in winter, a river drawn with squiggly lines. In some cases the red ochre of the lipstick-like stone pigment would be used to draw *pictographs*, and more painstakingly the ancient people would carve their images into the stone *petroglyphs*.

One remarkable possibility that is repeated throughout the world and particularly in the Americas in the prehistoric rock art is that of celestial events: possibly a supernova, bright comet, eclipse, or even a conspicuous constellation such as Orion which can represent a Hunter to these early people as in mythology today. Among the most prevalent images – with many variations – are what are thought to be illustrations of the SUN, usually in groupings of anthropomorphic and/or geometric figures.[3]

One such astronomical reference is thought to be of the brilliant supernova of 1054 AD which is depicted throughout the world as well as in the Americas, showing a crescent moon with a nearby second symbol depicted as a circle, a cross or a big asterisk. This remarkable event would surely have evoked not just curiosity but perhaps fear, as it was visible in daylight for 23 days and in the night sky for nearly two years. These symbols have been found at 12 sites in New Mexico, Arizona, Texas, California, Arkansas and Baja California. The culmination of all possible sites has been

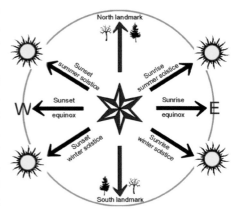

The observed points (azimuth 'Cardinal Directions') of the rising or setting sun throughout the year can be used accurately as an annual calendar to predict the onset of seasons, with the southernmost azimuth markers always occurring in winter, the northernmost indicative of summer. (P. Clay Sherrod)

3. Sherrod, P. Clay, *Motifs of Ancient Man*, University of Arkansas Office of Research in Science and Technology, 1980

studied by Dr. John Brandt with Dr. Stephen P. Moran of Goddard, and David L. Crawford of the Kitt Peak National Observatory in Arizona.

Other astronomical events depicted in the rock art archives are possibly constellation outlines of the Pleiades and Orion (Navajo) and illustrations of the grand Milky Way stars overhead, brilliant meteors and certainly comets (Hopi), all objects and events of great importance to the developing mind of the primitive ancestors.

Certainly more practical, much pre-Columbian rock art of the MesoAmerican cultures of Central America and Mexico used carved stone as intricate calendars and forecasters of the depth of winter and height of summer (solstices) and the coming of spring and the impending demands of fall (equinoxes).

Agriculture, Time-keeping and the Emergence of Pre-Columbian Astronomy

To discuss the evolution of stargazing and astronomical observations within the pre-Columbian cultures, we must look to the beginning of agriculture; for only through agriculture in any given location are the people given the precious resource of TIME in which innovative thought can progress. Also, with domestic agriculture comes the necessity of a calendar through which the passage of time can be calculated for the planting, nurturing and harvesting of crops, something very abstract and technical to prehistoric mankind.

As early as 10,000 years BP there is evidence of agriculture or harvesting in the Americas; archaeologists have dated ancient squash seeds, acorns, pine nuts and even grasses to Mesoamerican people and by 7,000 years BP it appears that corn was being developed for domestic harvest, and later came the sunflower and even cotton at about 4,600 years BP.[4]

When you domesticate crops for harvest, you must have some way to determine the solstices (winter and summer) and equinoxes (spring and fall); these timekeeping instruments can be in the form of markings in the Earth and rocks, or with calendars that could be set with the rising and setting of the sun at the points of rising or setting in the farthest northern horizons (summer) or farthest southern horizons (winter).

We divide the prehistoric cultures in the Americas into two segments: pre-agriculture and post-agriculture. As with any civilization in the world, the human mind advances exponentially, once the "stay-at-home" environment of growing

---

4. Ordish, George, et al., 'Origins of Agriculture', *Encylopaedia Britannica*, February 2020: **www.britannica.com/topic/agriculture/The-Nile-valley**

food emerges. Seven periods/cultures exist within these two segments, only one of which pre-dates agriculture both in the southern and northern Americas:

PALEOLITHIC – Nomadic Cave Dwellers – North American Rock Art – USA – Archaic period to 400 AD – oldest in Americas 14,800 years BP, non-agriculture.

OLMEC – began near Gulf of Mexico, western Tabasco, in the centuries before 1200 BC and declined around 400 BC.

AMERICAN MOUND BUILDERS – earliest 5400 BC in northern Louisiana

INCA – ANDES MOUNTAINS Peru – stoneworks as early as 400 BC – thrived 1400 to 1533 AD.

MAYA – CENTRAL AMERICA – Began in millennium before 1 AD; flourished by 300 AD and disappeared by 1100 AD.

AZTEC – MEXICO CITY – 1345 to 1533 AD.

AMERICAN TRIBAL – Anasazi, Hopi, Navajo, Cherokee, etc. – to present

As North American civilizations grew and prospered, so did the innovation and development of calendars and markers across the landscapes to predict the beginnings of seasons throughout the years by vantage points constructed high above the local terrain, many of these marked with great timbers to serve as celestial sighting posts for the rising and setting of the sun, moon, and even stars for the earliest priests. (Illinois Dept. of Conservation, Cahokia Museum)

Pre-Columbian Calendars

Halfway around the globe from our pre-Columbian civilizations, but perhaps contemporaneous with their development, Egyptian priests who served as the Pharaoh's astronomers would stand in a designated sacred area and observe the eastern horizon for the brilliant star SIRIUS to rise just ahead of the sun as the chill of spring would wane and the warmth of summer began. It was then that the brightest star in our skies could be seen against morning twilight that served as a signal that the Nile River would soon begin filling, flooding the fertile farmland in the delta below.

That date was critical for the huge population of the Egyptian empire, because the flooding meant that planting of crops was immediately essential so that they would be fully grown and harvestable prior to the long-lasting droughts of the Egyptian desert. In the valley of the Nile, life depended on it. Life depended on knowing a specific time of the year that signalled plentiful irrigation simply by the rising of a bright star.

Similarly, in the first century AD – perhaps even before – western pre-Columbian cultures were developing sophisticated calendars of their own using whatever creative resources might be available to them: stone, rock faces, natural landmarks, wooden poles and earthen mounds.[5] In this very concise summary on the highlights from scientists who study the astronomical importance in archaeology – *archeoastronomy* – the author will focus on two predominant technologies that were seemingly utilized in pre-Columbian cultures: *solstice markers* and *engineering alignments*. Indeed in many cases, the two technologies were inseparable and simultaneous in development. Of the thousands of suspected archaeological resources within the Americas which seem to possess some indication of the importance of the night sky, this overview can discuss only a few to perhaps put them in perspective both culturally and chronologically.

Often overlooked in such studies, even by archaeological scientists, is the need to consider precession of the Earth's axis of rotation over thousands of years; what are the sunrise and sunset azimuths of solstices and equinoxes today are not what those points were 4,000 or even 1,000 years ago; any study or assumption that does not incorporate the precessional shifts has not been properly investigated.

5. The concepts of celestial timekeeping and an overview of methods can be found in the Archaeological Conservancy's American Archaeology, Malakoff, David, 'Celestial Timekeeping', *American Archaeology* Volume 19, Number 1, 2015. Archaeological Conservancy), in which this author (Sherrod) discusses in that summary the importance of knowing that not every archaeological site is necessarily also a calendar or time keeping location, which sometimes is over-zealously assumed by researchers.

Solstice Markers

Prehistoric sites are constantly reported to archeoastronomers purporting to have potential "sun markers" – petroglyphs or pictographs which might somehow be used to indicate the point of rise or set of the sun at the solstices or equinox. Such is the tip that led to the remarkable discovery of the *Toltec Module* by Sherrod and Rollingson in 1982, a hitherto unknown technology of pre-Columbian cultures throughout the Mississippi valley, in which earthen mounds were constructed

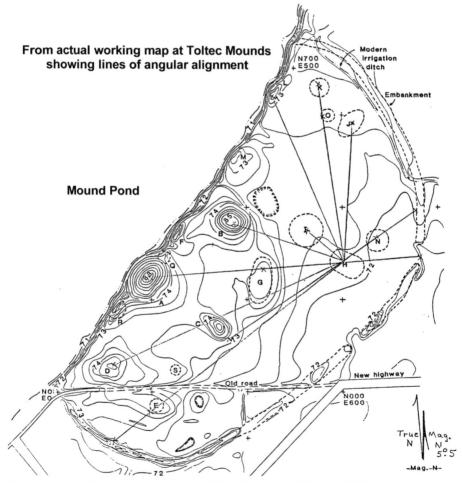

Demonstrating alignments and use of the Toltec Module in a Mississippi Valley mount site (with the concept utilized possibly throughout North American mound building cultures). (Sherrod and Rollingson, *Surveyors of the Prehistoric Mississippi Valley*, University of Arkansas/Arkansas Archeological Survey)

using both a standard "ruler" of scale and angles used to align to astronomical risings of the sun, moon and perhaps other celestial objects.[6]

Other markers are commonly found in stone rock faces in the form of "sun signs", rock art that can be spiral, circles with radiations, or other familiar line drawings that represent the importance of the sun to prehistoric cultures. There are certainly obviously depicted celestial times of great importance in the MesoAmerican influence, among them the Maya city of *Uxmal* in Mexico's Yucatan Peninsula. A prominent structure now known as the *Governor's Palace* appears to be aligned with another contemporaneous pyramid that can mark the azimuth where the planet Venus would rise once every eight years. There is a small level area on this pyramid which features a stone throne on which is carved a two-headed jaguar. Sitting in this "throne" affords a view of Venus rising over the nearby community of *Nohpat*, and subsequently setting over the palace at Uxmal on that same evening.

As will be discussed, this could be simply a coincidence, except that the palace is covered with glyphs known to represent Venus, a planet known to embody great importance to the Mayan people. Additionally, the *Dresden Codex*, a famous book of Maya astronomy dating it to 1200 or 1300 AD, demonstrates that VENUS played an important role in Maya cosmology and timekeeping. Several pages of the codex can be used to calculate Venus's movement across the sky, and then links the planet to important Maya deities, omens, and ceremonies. It is interesting to point out that Venus' movements served a very practical and necessary purpose: marking the beginnings of the region's dry and rainy seasons, an application similar to the Egyptian use of the first sign of the star SIRIUS rising to predict the flooding of the Nile River.

Structures, monuments and obelisks were not the only methods of timekeeping to the ancient Mayans; far more sophisticated were their development of elaborate calendars, such as *The Mayan Calendar* which nearly caused a world-wide panic in 2012. Although many of the functions of the calendar's numerous potential calculations have still not been deciphered, the scientific and technological – not to mention mathematical – input necessary for its conceptualization was outstanding, and remains outstanding until today.

The Mayan Calendar is only one of several variations that were developed and used through the course of this culture; most are defined as an "interlocking calendar", meaning that there are several moving components that must be set sequentially to actually ascertain its functions. It is based on cycles of 260 and

6. Sherrod, P. Clay and Martha Rollingson, *Surveyors of the Prehistoric Mississippi Valley.* Arkansas Archeological Survey Research Series #28, 1987

The basis of the Mayan Calendar: a system of interlocking symbols and wheels adapter from the Aztec SUN STONE (Piedra del Sol) depicting two cycles of time which ended mysteriously on the winter solstice 2012. (Museo Nacional de Antropología, Mexico City)

360 days, as well as a system for working through much longer periods by bundling days into groups of 20 and 13.

The short description of this device is beyond the scope of this overview; it is incredibly complicated. "The Maya were obsessed with time," notes Pomona College astrophysicist Bryan Penprase.[7] Indeed so obsessed that the world panicked at the prediction in 2012 of the END of the Mayan Calendar, which "ran out" on 21 December 2012 (interestingly the date of modern winter solstice). Started at a date of 11 August 3114 BC, the Mayan Calendar is quite specific about both the

---

7. Penprase, Bryan, *The Power of the Stars: How Celestial Observations Have Shaped Civilization*, Springer Publications 2010; **bryanpenprase.org/the-power-of-stars**

start and end dates and the method by which the long count of the calendar is computed, they did take into account the precession of the Earth.

What, exactly, was so notable or important with 3114 BC, the subsequent passage of 1,872,000 days and ultimately with the "end date" of 2012 AD? How did those ancient people possess the knowledge with which to computer precessional motions of the Earth through space? With complexity far too involved for discussion here, there are innumerable sources which can be accessed to explain perhaps the world's most sophisticated and complicated timepiece.[8]

We are left with an essential question that seems to have no answer in regard to the Mayan Calendar: WHAT is the significance of the staring date of 11 August 3114 BC – a date that is far before even the first Mayan construction or hint of cultural affiliation – and the ending date of 21 December 2012, a specific date for which there seems to be little or no correlation two thousand years in the future for those who design this intricate time piece?

Further Reading
An excellent resource for the continually growing collection of prehistoric rock art is the American Rock Art Archives **www.bradshawfoundation.com/america/index.php**

———

Part Two of this discussion, 'The Evolution of Ancient Skywatching as it Moves Northward in the Americas' will discuss sky watching, observations and use of the sky by cultures as they migrated northward in the western hemisphere and will appear in the *Yearbook of Astronomy 2024*.

---

8. See **en.wikipedia.org/wiki/2012_phenomenon** for an excellent discussion.

Tycho Brahe and the Parallax of Mars

David Harper

Introduction

Tycho Brahe (1546–1601) is celebrated as the greatest observational astronomer of the pre-telescopic era. He died just eight years before Galileo became the first person to look at the heavens through a telescope, but his life would be defined by another event which also changed the course of astronomy. Three years before Tycho was born, Nicolaus Copernicus died, and his book *De revolutionibus orbium coelestium* ("On the revolutions of the celestial spheres") was published posthumously. This was the book which re-introduced the heliocentric model of the solar system to European astronomy, challenging the Earth-centred model that had been advocated by Aristotle and Ptolemy in ancient times and which remained the unquestioned truth as far as most astronomers – and the Church – were concerned.

The Earth Cannot Move

For more than a century after its publication, many astronomers were reluctant to agree that *De revolutionibus* described the solar system as it actually was. They preferred instead to regard it

A contemporary portrait of Tycho Brahe by the Dutch artist Jacob de Gheyn II. The caption (in Latin) dates the engraving to 1586 when Tycho was 40 years old. The engraving also bears Tycho's motto, NON HABERE SED ESSE (meaning *not to seem, but to be*). (Wikimedia Commons/Jacques de Gheyn II/Museum of Fine Arts, Houston)

as a convenient mathematical artifice which happened to simplify the calculation of the positions of the planets. This was Tycho's own position. His objections were not primarily religious. There were valid philosophical and scientific reasons for thinking that the Earth could not move. For one thing, the stars showed no measurable parallax over the course of a year. If the Earth was indeed orbiting the Sun, the stars should appear to move back and forth. Tycho had the most accurate

An aerial photograph of the island of Hven, the site of Tycho's observatories Uraniborg and Stjerneborg. It lies in the Øresund strait between Denmark and Sweden. Now named Ven, the island was ceded to Sweden in 1660. (Wikipedia/Bjørn Christian Tørrissen)

instruments ever constructed; capable of measuring the positions of the stars to within one arc-minute, and even he could detect no annual parallactic movement of the stars.

With the benefit of four centuries of hindsight, we know that Tycho could not possibly have detected the parallaxes of the stars with his instruments, excellent though they were. The parallax of even the nearest star, Proxima Centauri, is less than one arc-second, and no astronomer would accomplish such a feat of observation until Friedrich Bessel measured the parallax of 61 Cygni in 1838. But astronomers of Tycho's era were unaware of the true distances to the stars, and thought that they were relatively close. Indeed, Tycho himself believed that he had measured the apparent diameters of the brightest stars – one arc-minute for third-magnitude stars, and up to 3 arc-minutes for the brightest first-magnitude stars.

The Copernican model also required the Earth to spin on its axis to cause the rising and setting of the Sun and stars. If it did this, then it stood to reason that a cannonball fired due south would veer eastwards and miss its target, and a lead weight dropped from a tall tower would also veer and land at a distance from the foot of the tower. Neither of these phenomena was observed to happen, so the Earth could not be rotating.

Another strong argument came from the long-held belief that the Sun and planets were made from an ethereal form of matter whose natural state was rapid movement. The Earth, by contrast, was heavy and dense, and it was absurd to imagine that it could move.

The Parallax of Mars

Tycho realised that it was possible to test the Copernican model directly by observing Mars. In the prevailing geocentric model, Mars always remained further from the Earth than the Earth-Sun distance. According to Copernicus, however, Mars can approach the Earth much more closely at opposition. In modern terms, the Earth-Mars distance can be much less than one astronomical unit in the Copernican system. This means that the diurnal parallax of Mars should be larger than the diurnal parallax of the Sun.

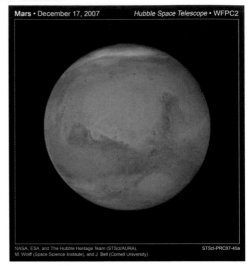

The Hubble Space Telescope captured this stunning image of Mars a day before its closest approach to Earth in December 2007. (NASA/ESA, the Hubble Heritage Team (STScI/AURA), J. Bell (Cornell University), and M. Wolff (Space Science Institute, Boulder))

The diurnal parallax is the difference in the apparent position of an object seen simultaneously by observers on opposite sides of the Earth. We now know that the Sun's diurnal parallax is a little less than 9 arc-seconds, but in Tycho's time, it was believed to be 2′ 50″, a value handed down from Ptolemy. In the Copernican system, Mars should have a diurnal parallax twice than of the Sun, amounting to five or six arc-minutes. Tycho knew that his instruments could easily measure such a large angle.

His method was simple and elegant. He would observe Mars on nights close to opposition, when it was visible for most of the night. Using his state-of-the-art sextants and quadrants, he would measure the angular distance between Mars and selected stars soon after sunset, and again before dawn. Measurements taken at the same time on successive nights allowed him to calculate the movement of Mars caused by its orbital motion, which he could then subtract from his observations. The difference that remained would yield the diurnal parallax.

He applied his method for the first time during the winter of 1582/3 when Mars was at opposition among the stars of Gemini. He was unable to detect a measurable parallax, which appeared to vindicate Ptolemy's geocentric model.

Mars came to opposition again early in 1585 in Leo, and Tycho and his assistants observed it carefully, measuring its distance from selected stars at the start and end of each night. When he analysed the observations, he got an unwelcome surprise. There was a parallax, but it was negative, which is impossible!

A Problem with Refraction

Tycho was aware that the Sun and planets always appeared slightly higher in the sky than they should, and that the effect was greater at low altitudes, amounting to several minutes of arc. We know this as refraction, but in Tycho's day, it was poorly understood. Alas, Tycho's method for measuring the parallax of Mars often required him to observe the planet when it was close to the horizon, and he had no way to correct his observations for the effects of refraction. He recognised that this was the most likely explanation for the negative parallax of 1585, and the following year, he made detailed observations of the altitude of the Sun during the course of a day in June. From these, he constructed a table of refraction which agrees very closely with values calculated by the best modern formulae.

Unfortunately, this was not the refraction table which he used to analyse the observations which he made at the 1587 opposition. The underlying cause of refraction was not understood, and Tycho believed that the Sun, stars and planets were each affected in different ways. Rather than using his excellent table of solar refraction, he constructed a new table specifically for Mars, based partly on theory. It over-estimated refraction at some altitudes and under-estimated it at others.

The Tychonic System

The 1587 campaign was the most detailed which Tycho had yet undertaken. Mars came to opposition in March of that year among the stars of Virgo. In addition to his trusted sextant and quadrant, he introduced an equatorial armillary sphere which allowed him to measure right ascensions and declinations directly.

Using his flawed refraction table, he analysed the observations and found – at last! – a parallax for Mars. This was a cause for celebration, because by this time, Tycho had devised his own model of the solar system which combined elements of both the Ptolemaic and Copernican systems.

In fact, Tycho's system was simply the Copernican system, but with the Earth fixed in space rather than the Sun.

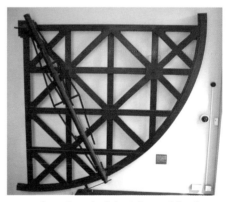

A mural quadrant built by John Bird for the Radcliffe Observatory, Oxford, in the late 1770s, and now part of the collection of the Museum of the History of Science, Oxford University. This instrument uses a telescope for sighting. Tycho's quadrants, two centuries earlier, pre-dated the telescopic era and would have used slits at either end of a sighting rod. (Wikimedia Commons/Heinz-Josef Lücking)

It explained the observed movements of the planets, including the troublesome retrograde motion of Mars, Jupiter and Saturn, as effectively as Copernicus had done, but without requiring the convoluted epicycles of Ptolemy, or the philosophically unacceptable and scientifically absurd movement of the Earth proposed by Copernicus.

The discovery of a parallax for Mars supported Tycho's model as well as that of Copernicus, since Mars approached the Earth more closely than the Sun in both systems.

The Later Years

When Mars returned to opposition in late April 1589, it was in Libra, and significantly further south than in 1587. At Tycho's observatory, almost 56° N, it culminated barely 23° above the horizon and remained visible for only a few hours each night, even at opposition. It was no longer possible to make observations that would yield a parallax. These unfavourable conditions continued whilst Mars came to opposition in the summers of 1591 and 1593.

At the opposition of November 1595, Mars was once more well north, among the stars of Taurus. Tycho's records show that his assistants made numerous observations, but there is no evidence of the calculations that would have yielded a parallax. It seems that Tycho had lost interest in the parallax of Mars. His personal situation had also deteriorated by this time. The young king Christian IV was more interested in military conquest than science, and failed to honour his father's promise to allow Tycho's heirs to inherit the castle of Hven. Tycho left the island, and his observatory, in 1597. He would never return. He died in exile in Prague in 1601.

And Yet It Moves

As is well known, Tycho's observations of Mars, made to disprove the heliocentric system of Copernicus, were later used by Johannes Kepler to derive his three laws of planetary motion. Kepler's laws, in turn, were shown by Isaac Newton to be the natural consequence of planets orbiting the Sun under a force which varies as the inverse square of the distance.

Indirectly, then, Tycho's remarkable series of measurements of the precise position of Mars helped to demonstrate that the Copernican system is indeed a true representation of the solar system.

James Bradley, the third Astronomer Royal, provided the first direct observational evidence that the Earth orbits the Sun. In January 1729, he reported to the Royal Society that he had observed an annual periodic variation of 19.5 arc-seconds in the

The crater Tycho is a prominent feature of the Moon's southern highlands. Samples of its bright ray system obtained by the Apollo 17 astronauts suggest that it is 108 million years old, making it one of the youngest features on the Moon. (Wikimedia Commons/NASA/James Stuby)

declination of the star Gamma (γ) Draconis which could only be explained by the motion of the Earth around the Sun combined with a finite speed of light. This was Bradley's celebrated discovery of the aberration of starlight.

Tycho's Legacy

Most brief summaries of the life of Tycho Brahe focus on his skill as an observer, the supernova of 1572, and Kepler's later use of his observations. They may also mention his prosthetic nose or the dwarf who was supposedly part of his retinue at the castle of Hven. But Tycho also deserves an honoured place among theoretical cosmologists.

Further Reading

Blair, A. (1990) 'Tycho Brahe's critique of Copernicus and the Copernican system.' *Journal of the History of Ideas* 51:355-377 (DOI: 10.2307/2709620)

Gingerich, O., Voelkel, J.R. (1998) 'Tycho Brahe's Copernican Campaign.' *Journal for the History of Astronomy* 29:1-34 (DOI: 10.1177/002182869802900101)

The early chapters of Allan Chapman's book *Stargazers: Copernicus, Galileo, the Telescope and the Church* (Lion Books, 2014) also provide a fascinating insight into the life of Tycho Brahe.

A Tale of Two Henrys and the Search for Their Great Telescopes

Gary Yule

In June 2019 I was contacted by Brian Jones who asked me if I would take up the challenge of trying to locate the current whereabouts of Henry Lawson's eleven-foot George Dollond refractor, and write an article about my search for the *Yearbook of Astronomy*. Lawson was not an astronomer I had heard of so, with the honour of being part of the Yearbook and the chance to do some digging into astronomical history, I duly took up the challenge!

Henry Lawson

My search began with looking into Henry Lawson himself, who was born at Greenwich on 23 March 1774. He was the second son of John Lawson, Dean of Battle, Sussex, and may have been a descendant of Catherine of Parr (Queen of England 1543-47 and wife to Henry VIII). Evidence of this is mentioned in his obituary.[1]

Henry Lawson. (Wikimedia Commons/ Wellcome Library, London/ W. Giller/ C. Lucy)

He became an apprentice at an optical establishment owned by his stepfather, Edward Nairn of Cornhill. After completing his apprenticeship he went on to become a Freeman of The Worshipful Company of Spectacle Makers in 1803–4 and 1822–23. Following his marriage in 1823, he moved to Hereford where he built an observatory, equipping it with a five-foot refractor in 1826 and one of eleven feet in 1834, the latter made by George Dollond and reputed to be one the finest instruments of its time.

---

1. Henry Lawson's obituary can be found in *Monthly Notices of the Royal Astronomical Society*, Vol. 16, p.73 **articles.adsabs.harvard.edu/full/seri/MNRAS/0016/0000086.000.html**

In 1841 Henry moved to Bath where he took up residence at 7 Lansdown Crescent (part of a crescent of houses constructed between 1789 and 1793 on Lansdown Hill, from where they presented a clear and unobstructed view over central Bath). He converted the top floor of the property into an observatory where he installed a wonderful example of an equatorial platform, the design for which he never gained the credit he deserved. The platform was fixed from the roof to the floor of the room to mount the eleven-foot refractor. He was recognized for his design of the recliner chair, which he originally designed to facilitate astronomical observing, but went on to be used for a range of other purposes such as wheelchairs for the elderly and infirm.

Lawson installed his telescopes in the observatory, the eleven-foot focal length Dollond refractor being placed on the equatorial mounting. This fine instrument was equipped with a double screw cobweb micrometer, the webs of which were illuminated by a lamp placed at the side of the telescope, the equivalent of an illuminated reticule today. It also incorporated a spherical double refracting crystal micrometer in combination with a double refracting prism, to increase the angle of separation of the sphere. The telescope had an achromatic lens, and nine Huygenian eyepieces, giving a range of possible magnifications up to the power of 1400x.

Lawson went on to make many observations from his house in Bath, including an occultation of Saturn (8 May 1832) and of the comet discovered on 3 December 1839 by Johann Gottfried Galle (December 1839 and January 1840). He also

In this image, taken from Lawson's 'A Paper on the Arrangement of an Observatory for Practical Astronomy & Meteorology' (1844), we see the top floor of Lawson's house at Lansdown Crescent, Bath, housing the equatorial mount, eleven-foot refractor and recliner. (Alamy Stock Photo)

288   Yearbook of Astronomy 2022

recorded many meteor and solar observations, including the annular solar eclipse of 15 May 1836.

He became a Fellow of the Royal Astronomical Society in 1833, the Royal Society in 1840, and the British Meteorological Society in 1850, and went on to publish two articles: 'A Paper on an Arrangement of an Observatory for Practical Astronomy and Meteorology' (1846) and 'A Brief History of the New Planets' (1847).

In December 1851 Lawson proposed to donate his astronomical and meteorological instruments, together with a large sum of money, to the Town and County of Nottingham, on the proviso that funds could be raised to build an observatory. A committee was founded, with the Duke of Newcastle as Chairman; the committee relied on 727 subscriptions and money granted by the Government, along with Lawson's donation and value of the equipment. Unfortunately, there was some sort of dispute as to the actual value of the equipment, and the proposal never came to fruition.

Henry Lawson died at Bath on 22 August 1855. According to his obituary, in the year before he passed he presented the eleven-foot Dollond refractor and attached equipment to the Royal Naval School at Greenwich.

The Search Begins

The Royal Naval School became the starting point for my search. I made contact and was pointed in the direction of Simon Marsh (Director of Development) at the Royal Hospital School. Simon informed me that the original site of the Hospital/ Naval schools had changed many artefacts and records had been moved around or placed in storage. He passed me on to Rebecca Young (Archivist) at the Royal Hospital School, who was very helpful but informed me that they had no record of Henry Lawson or a Dollond refractor of this size, thus drawing a blank.

My next step was to contact the Royal Astronomical Society and Royal Greenwich Museum. I made contact with Dr. Sian Prosser (Librarian and Archivist for the RAS). Dr. Prosser was very helpful and carried out a search of the RAS archives for details of Lawson, returning with a multitude of links, including his obituary and records of his solar observations which are still in the RAS archives.

Sian also informed me that the telescope was actually presented to the Royal Naval School, not the Hospital School. The Royal Naval School was based at New Cross, South London at the time of Lawson's death in 1855. It closed in 1910 and became a charity, its records being transferred to the Royal Naval College. The purpose-built school at New Cross is now occupied by the Goldsmith College, University of London, and the Hospital School's premises in Eltham now house Eltham College.

This left me with three possible sources, these being the records of the Royal Naval School (which were transferred to the Royal Naval College and are now held at the National Archives); the records of the Goldsmith College; and the archives of Eltham School. I followed up on all three to see if there was any record of the observatory or equipment, although my search came to a dead-end, as I received no reply from either the Goldsmith College or Eltham School. Neither did I come across any mention of Lawson, his observatory or equipment in the records that I could access online. I even tried aerial reconnaissance photographs of the area and grounds of both sites, but to no avail.

Luckily, at this time I was carrying out some research into an 1826 William Cary of London equatorially mounted refractor held at the Chaseley Field Observatory in Salford. I had contacted an old member of our society Christopher Lord, who has a keen interest in old optics. When I mentioned Lawson's refractor he remembered hearing something about this telescope in the past. Chris informed me that, during the time when the Hospital School and Royal Naval School shared the same site, a lot of the records were documented incorrectly with some donations to the schools being indexed under the wrong titles. I was also pointed in the direction of the Survey of Astronomical History[3] and the Lewisham Local History archives.

This image, taken from the book *The Royal Hospital School Greenwich* by H.D.T. Turner, shows the three cylindrical domes on the observatory at the Royal Hospital School's original site at Greenwich, one of which is said to have housed Lawson's refractor. (H.D.T. Turner/Phillimore & Co. Ltd)[3]

2. There are no acknowledgments to any of the images appearing in the book and all efforts made to locate the copyright holder were unsuccessful.

3. For more information on the Survey of Astronomical History visit **shasurvey.wordpress. com**

At Last Some Evidence

The Survey of Astronomical History revealed:

- Royal Hospital School, Greenwich (1849/59–1930s) in operation by 1860, with instruments by Gilbert donated by the admiralty that were made The East India Company – St Helena Observatory.
- The building of Portland stone was originally fitted with two cylindrical domes but later extended with a larger thirteen-foot dome that housed an eleven-foot refractor. This telescope was originally owned by Henry Lawson.
- Fisher reported observations of the 1860 solar eclipse from the observatory using Lawson's telescope.

On presenting the information to the Royal Hospital School, we still drew a blank as to the whereabouts of the actual telescope, and it seems it may have been one that ended up in a private collection. However, Christopher Lord gave me the name of someone that many readers of this article may know – Ron Irving. Ron was a great collector of antique telescopes and Huygenian eyepieces. It seems after the Second World War, many observatories and telescopes were left to decay, as the money to fund their upkeep was generally no longer available. A great number of observatories and their equipment were put out for scrap. Ron took it upon himself to collect the disowned equipment and eventually accumulated quite a collection. Sadly he is no longer with us, and his collection has since been dispersed or lost along the way.

My search for Henry Lawson's eleven-foot George Dollond refractor may have drawn a blank, although delving into the past and being introduced to the people I have met along the way has been well worth the effort. However, if anyone out there has any information on Lawson's telescope, please feel free to get in touch.

Henry Brinton

As the above story ended in disappointment, I thought I would go in search of another great historical telescope, the twelve-inch Newtonian reflector owned by Henry Brinton. In view of the fact that this instrument and its owner featured on the front cover of the *Yearbook of Astronomy 1963*, it seemed fitting to try and locate its current whereabouts, and provide some information on Brinton himself.

Henry A. Brinton was born in Wolverhampton on 27 July 1901. At an early stage in his career, Brinton was mainly involved in political circles. He stood at several by-elections on behalf of the Labour Party, his view on politics soon changed and before 1939 he was an active member of the League of Nations Union, where he spent some time in Spain. Apparently, at one point there was a price placed upon his head, by direct order of General Franco!

On returning home, he was the main organizer of a large refugee camp for Basque children. He gave up politics in 1955 but kept a hand in public life as Chairman of The Regional Hospital Board and pioneered for better standards of education for Sothern State Schools. Brinton was also a keen writer and published thrillers under his name and the pseudonym of Alex Frazer.

Henry A. Brinton. (Alicia Giambarresi)

His interest in astronomy had been long-standing, although it wasn't until he moved to Selsey, Sussex in 1957 that he joined the British Astronomical Association (BAA), going on to play an active role and serving on the council for many years.

He equipped himself with a fine twelve-inch Truss Tube Newtonian Reflector on an equatorial mounting, this instrument formerly belonging to Robert Barker, a well-known member of the BAA. Henry also had an interest in radio astronomy

Brinton and a film crew at Selsey in 1971 during the making of an episode of The Sky at Night. (Alicia Giambarresi)

and built a radio telescope in the grounds of the house at Selsey.

Brinton became an avid astrophotographer and produced some excellent images of the Moon and planets, some of which can still be found in his books including *Astronomy for Beginners* and *Measuring the Universe*. He also made several appearances on the Sky at Night and became very close friends with Patrick Moore. In notes from his daughter, Alicia Giambarresi in the BAA journals, apparently therefore Patrick moved to Selsey as Henry and Patrick had ideas of sharing equipment for research. It was Alicia who found Farthings for Patrick during a drive through the village, the house not being listed with any of the local real estate agents. Henry lived at the Old Mill House in Selsey with his

Brinton's telescope pictured at Bayfordbury Observatory, Hertfordshire. (Flickr/University of Hertfordshire/Bob Forrest)

wife Ailee and two adopted daughters Alicia and Julie. Sadly the house no longer stands, and the site has been developed into six cottages.

In 1975 Briton suffered from the first of his strokes, losing the use of his right arm. Realizing he could no longer use his equipment, he made the generous gesture of donating it all to the Hatfield Polytechnic Observatory, now the University of Hertfordshire, Bayfordbury Observatory, in 1976. The telescope was housed at Bayfordbury for several years where it was used for outreach and basic training of students in using a telescope. In 2009 it was relocated to Slindon College near Arundel, Sussex where it stands today. Henry Brinton passed away on 1 June 1977.

Mission to Mars
Countdown to Building a Brave New World
Laying the Foundations

Martin Braddock

Introduction

The last article entitled 'It All Starts With a Journey' in *Yearbook of Astronomy 2021* presented some obstacles for gaining a greater understanding of the space environment and some steps that we will need to take to enable astronauts to live and work effectively and safely on the journey to Mars. We will now focus our attention on laying the foundations for the colonisation of Mars and what investigational considerations and physical work will need to be conducted prior to humans landing on the planet.

With the closest distance between Earth and Mars being around 55 million kilometres, transporting large volumes and masses of equipment and materials this far would be logistically very challenging by today's technology and would involve an armada of costly space missions, having their own risks for success. It is estimated to cost approximately $4,000 to launch a kilogram of material out of Earth orbit, let alone transport it to Mars and approximately nine times the weight of propellant is required per kilogram of cargo for launch alone, making it simply not financially viable to transport large volumes of building materials such a great distance. This situation provides scientists with both technical challenges and opportunities, and drives the current set of planning assumptions that laying the foundations ahead of any human presence on Mars will be critical to the success of inhabitation and future colonisation. Indeed, space colonisation, at least in the first instance, will likely have the way paved by fully automated processes (Campo et al 2019).

Make the Best of What You Have

Although unprecedented, there are some parallels with many Earth bound expeditions of previous centuries (Steward and Greenwood 2016). Some historical examples include Christopher Columbus' first voyage to the West Indies in 1492, Vasco da Gama sailing to India in 1497, James Cook's voyage to discover Australia in 1768 and, perhaps the most important prior to the Moon landing in 1969,

Ferdinand Magellan's first circumnavigation of the Earth in the 1520s. In all of these cases some supplies were taken to sustain the expedition during the initial phases though longer term success depended upon securing local resources. For Martian exploration and settlement to be achievable on a sustainable scale, it will be necessary to use materials found on Mars and to develop technologies ahead of landing that maximise our ability to pre-fabricate as many building blocks for settlement as possible, conducted entirely robotically. Achieving this aim is enabled by in-situ resource utilisation (ISRU) technologies, and space agencies and private enterprises plan to send robotic 'construction workers' in advance of human crews to harvest raw Martian materials and process them into parts using 3D printers which can be assembled into habitats and other structures to provide shelter or to grow food. Using technologies developed and first tested on Earth, this will ensure that our future Mars habitats are built ahead of our arrival and function correctly from day one.

What is the Martian Soil Made Of?

The Martian 'soil', more correctly named regolith, is a dusty, pulverised rock layer which has been produced by the impact of asteroid and meteor collisions with the planet's surface together with the erosion of iron-rich igneous rock by physical weathering over billions of years. As is the case with builder's concrete on Earth, when mixed with water it should form a cement which may be used to bind surface rocks or rocks mined from the sub-surface together to make pre-fabricated building blocks. A technology to extract water from regolith has also recently been developed (Trunek et al 2018).

As no craft has landed on Mars and returned to Earth with samples of regolith for study, how can we even think of constructing buildings safe for human habitation with 3D printers? The answer comes from using regolith simulants, replicas of the real thing. Previous expeditions to Mars such as the Viking, Pathfinder,

Let us put our robotic construction crews to work. All the material to work with to build our new colony is all ready to be collected! (PIXABAY)

Spirit, Opportunity and Curiosity landers have analysed the chemical composition of regolith and found it to be made up mainly of silicon, iron, aluminium, magnesium and calcium oxides. The first simulant called JSC Mars-1 was developed in 1997 and a brief history of regolith development is shown in the table.

Date	Regolith	Location	Source details
1997	JSC Mars-1	Pu'u Nene cinder cone, Hawaii, USA	Basaltic lava
2007	MMS-1	Saddleback mountain, Mojave desert, USA	Iron-rich basaltic lava
2009	Salten Skov I	Mid-Jutland, Denmark	Mainly crystalline iron oxides
2011	ES-X	United Kingdom	ES-1 (dust), ES-2 (aeolian sand), ES-3 (coarse sand)
2015	JMSS-1	Hannuoba field, North China craton and Hebei province, China	Basaltic lava supplemented with magnetite and hematite
2020	NEU Mars-1	Chahar volcanic group, Wulanchabu, China	Basaltic lava

Table showing date of publication, name location and source details for selected Martian regolith[1] simulants. JSC = Johnson Space Center; MMS = Mojave Mars simulant; JMSS Jining Martian soil simulant; ES = Engineering soil. (M. Braddock).

Since then, NASA, the European Space Agency, the Chinese Academy of Sciences and Universities in China and Slovakia have developed several different types of simulated regoliths producing faithful replicas when comparing physical, chemical, mechanical and electrical properties. The most recent simulant NEU Mars-1 compares very favourably with the properties of previously described simulants (Guan et al 2020).[1] It is gratifying that scientists are developing regolith simulants of slightly different geophysical properties as the composition will likely vary for each location on the planet's surface due to variability in asteroid and meteor collisions and weathering by wind and water in ancient oceans and river structures. In addition, as we will not know how the simulants behave in a 3D printer until an automated device commences manufacture on Mars, ensuring robust and reproducible fabrication of structures using foundation materials of slightly differing properties should reduce the overall risk and increase the probability of success.

---

1. For further information on details of planetary regolith simulant visit **simulantdb.com**

Getting the 3D Printers to Do the Work

The rise of crowd-sourcing and subsequently crowd-funding to secure a diverse range of imaginative, inventive and practical proposals, together with a mechanism to reduce risk by having multiple investors, is both a plausible and accepted business model for getting a new venture established. One such concept is the generation of structures from Martian regolith using 3D printing technology dedicated purely to ISRU. 3D printing is now a well-established technology used on Earth[2, 3] though full integration into the supply chain requires further consolidation (Chan et al 2018). The challenge in using Martian regolith is to develop the technology with source material that can consistently produce stable pre-fabricated structures, with reproducible hardening times and with strength properties able to withstand the Martian weather. In such systems using 3D printing, complex and large structures can be built by depositing building material layer on layer for each part to a pre-specified design. One of many examples comes from National Aeronautics and Space Administration (NASA) who initiated a 3D Printed Habitat Challenge in a crowd-funding model[4] where scientists were asked to submit ideas for design architecture and then to build some prototype structures from simulated regolith material. In May 2019, AI SpaceFactory was announced as the winner of this crowd-sourcing event, having made a 4.5 metre tall prototype habitat constructed from a degradable, recyclable biopolymer basalt fibre composite derived from natural material regolith simulants.[5]

To further prepare for the future, engineers at NASA's Johnson Space Center and the Georgia Institute of Technology have designed and fabricated a functional model Mars rover using 3D printing.[6] The model, based upon the Resource Prospector 15 (RP15) prototype from 2015, uses 12 different motors to drive its four wheels and is capable of manoeuvring around obstacles. It is believed that once the engineering problems are solved and the process is scaled, these robotic rovers

2. The Positive Effects of 3D Printing Accessibility on Society **www.machinedesign. com/3d-printing-cad/article/21837540/the-positive-effects-of-3d-printing-accessibility-on-society**
3. The State of 3D Printing 2020 Edition. **Sculpteo.com**
4. STMD: Centennial Challenges **www.nasa.gov/directorates/spacetech/centennial_ challenges/3DPHab/index.html**
5. AI SpaceFactory wins NASA's 3D Printed Habitat Challenge **aispacefactory.com/post/ ai-spacefactory-wins-1st-place-prize-for-nasa-3d-printing-habitat-challenge**
6. NASA Engineers Use 3D Printers to Build a Mini Mars Rover **www.techeblog.com/ nasa-3d-printed-mini-mars-rover**

Custom built houses all ready to welcome us – different and beautiful! (AI SpaceFactory)

could be adapted to move large quantities of surface rock and regolith to fuel the pipeline of awaiting 3D printers for habitat construction.

Other crowd-sourcing events[7] have stimulated thinking about developing new technologies, such as in a Martian garden design in Switzerland to test a new imager, building a simulated Mars colony in the Emirati Desert, constructing bricks using pile driving compression techniques and making an ice house on Mars!

As for future building of structures with simulated regoliths, studies conducted at Northwestern University and the University of Berlin have shown that simple processes can be applied towards the fabrication of elastic structures (Jakus et al 2017) and stable ceramics (Karl et al 2018). These early studies illustrate the potential of this technology to be applied to the manufacture of hard landscaping materials suitable to withstand the Martian environment and this remains an intense area of scientific research today.

In summary, we believe that we are rapidly developing the skills and technology to solve many problems on Earth, including meeting the need for rapid construction

7. Six Projects Solving the Problem of Building on Mars. **constructible.trimble.com/ construction-industry/spacex-to-mars-city-how-to-build-on-mars**

Mars rover – perhaps future working versions will be 3D printed in situ? (PIXABAY)

of affordable housing[8] which can be recycled, and the efforts of humankind will have a profoundly positive benefit for the construction of cost-effective, viable habitats on Mars. Laying the foundations for the first wave and future generations of colonists will likely be from a workforce of automated construction workers using pre-fabricated parts produced in situ from a factory of 3D printing machines. The goal will be to use as much material from the Martian surface and subsurface as possible which reinforces the continued need for surveying the planet to discover new sources and perhaps new materials for use in the future.

The next article in this series will address habitat design in more detail and discuss what protection could be provided to astronauts to safeguard their physical and mental well-being. Planet Earth's recent COVID-19 pandemic teaches us much about human behaviour and co-operation, and we will explore some of the basic issues surrounding sustainable societal build in the next instalment.

---

8. What is your challenge? **Jacobs.com/what-if**

References

Campa, R., Szocik, K., and Braddock, M. 'Why space colonization will be fully automated' (2019). *Technological Forecasting and Social Change* **doi.org/10.1016/j.techfore.2019.03.021**

Chan, H. K., Griffin, J., Lim, J. J., Zeng, F., and Chiu, A. S. F. (2018). The impact of 3D Printing Technology on the supply chain: Manufacturing and legal perspectives. *International Journal of Production Economics*, 205, 156–162

Guan, J.-Z., Liu, A.-M., Xie, K.-Y., Shi, Z.-N. and Kubikova, B. (2020). 'Preparation and characterisation of Martian soil simulant NEU Mars-1'. *Transactions of Nonferrous Metals Society of China* 30: 212–222

Jakus, A.E., Koube, K.D., Geisendorfer, N.R. & Shah, R.N (2017). 'Robust and Elastic Lunar and Martian Structures from 3D-printed Regolith Inks'. *Scientific Reports* 7:44931

Karl, D., Kamutzi, F., Zocca, A., Goerke, O., Guenster, J. and Gurlo, A. (2018). Towards the colonisation of Mars by in-situ resource utilization: slip cast ceramics from Martian soil simulant. *PLOS ONE* 13: e0204025

Steward, M. and Greenwood, A. (2016). *Great Expeditions : 50 Journeys that Changed our World*. HarperCollins Publishers. ISBN: 978-000819629-5

Trunek, A.J., Linne, D.L., Kleinhenz, J.E. & Bauman, S.W. (2018). 'Extraction of Water from Martian Regolith Simulant via Open Reactor Concept'. NASA **ntrs.nasa.gov/archive/nasa/casi.ntrs.nasa.gov/20180002377.pdf**

Ad Astra: A Personal Journey

David H. Levy

The most impressive thing about the night sky, at least to me, is that it is not just there; it is changing. The Moon appears, or does not appear, in a radically different spot each night. The planets crawl smoothly along the ecliptic, and as our own Earth moves around the Sun, our home star appears to move along in response. Shooting stars, better known as meteors, scratch the sky at odd intervals, and at rare intervals, a comet will stretch its ghostly tail across a portion of the heavens. Even more rarely, a nova, an exploding star, will brighten the night.

I have been passionate about the night sky since I was a shy youngster of ten, away at a summer camp in Vermont. On the evening of 4 July 1956, we were watching fireworks to celebrate the Fourth of July American holiday. Our youngest cabin was sent to bed near the end of twilight. As we walked up a hill to the cabin, I looked skyward and saw a shooting star. It wasn't very bright, perhaps second magnitude. It appeared out of Draco and was headed towards a bright star that I now recognize as Vega. That same summer, while lying on my cot in the same

A Lyrid meteor imaged from Castor House, Jarnac Observatory on 21 April 2020. (David H. Levy)

cabin, I peered out of the screen and saw some stars that shone like faint beacons on that lonely night. Could these be the same stars that shone over our home? If they were, then possibly home was not so far away.

As a shy youngster I had few friends, and I was also susceptible to frequent depressions. I still am, but whereas earlier in my life depressions could last for weeks or months, now they evaporate after hours. Somewhere around that time, I thought that I could make friends with some of the stars. These friends would never abandon me, and they would always be there. One day, while visiting home during a long hospitalization for Asthma at the Jewish National Home for Asthmatic Children, Dad woke up from a nap in a terrible mood, and he accused me of "doing nothing but looking at his damned stars." His words upset me at the time, but over the years he rethought his attack and came to enjoy the "damned stars" with me on some nights.

My wife Wendee and I standing in front of one of our telescopes. Wendee is a co-discoverer of the periodic comet P/2010 E2 Jarnac (see below). (Joan-ellen Rosenthal)

Reading and the Stars

There is one other important aspect to the dawn of my interest in astronomy. Mom once criticized the fact that both my younger brother and I hated to read. I took her advice seriously and began to read book after book. Years later, Dad chimed in with a recommendation that he would like me to inherit his love of Shakespeare. I followed that, but for a long time I failed to see any connection between Shakespeare and the sky despite it having been displayed in front of me:

> *When beggars die, there are no comets seen;*
> *The heavens themselves blaze forth the death of princes.*
>
> (Julius Caesar 2.2.30–31)

The realization that the sky and English literature were connected did not hit me until a mild spring evening in late April, 1976. We were enjoying a Lyrid meteor shower and we saw several meteors. As I watched, I let my mind wander. How many other sky watchers have viewed this same meteor shower, whose meteors emanate from Comet Thatcher, in ages past. That comet passed us by in the spring of 1861. As I watched, my mind wandered further. Is it possible that non-astronomers, even poets, watched these meteors once upon a time?

The very next day I travelled to Kingston, in the Canadian province of Ontario, and I met Dr. Norman McKenzie. I told him about the previous night's session, and right away he recommended the poetry of Gerard Manley Hopkins. I knew Hopkins from high school, and remembered him as one of the most impossible poets to read in all of English literature. *The Windhover*, with its complex rhythm and dappled, dawn and drawn vocabulary, was not something that was attractive to me. But as I read on, I stumbled across a poem Hopkins wrote while he was attending Balliol College at Oxford. *I am like a slip of comet* was simple, elegant, and absolutely lovely.

> – *I am like a slip of comet,*
> *Scarce worth discovery, in some corner seen*
> *Bridging the slender difference of two stars,*
> *Come out of space, or suddenly engender'd*
> *By heady elements, for no man knows;*
> *But when she sights the sun she grows and sizes*
> *And spins her skirts out, while her central star*
> *Shakes its cocooning mists; and so she comes*
> *To fields of light; millions of travelling rays*
> *Pierce her; she hangs upon the flame-cased sun,*
> *And sucks the light as full as Gideon's fleece:*
> *But then her tether calls her; she falls off,*
> *And as she dwindles shreds her smock of gold*
> *amidst the sistering planets, till she comes*
> *To single Saturn, last and solitary;*
> *And then goes out into the cavernous dark.*
> *So I go out: my little sweet is done:*
> *I have drawn heat from this contagious sun:*
> *To not ungentle death now forth I run."*

Years later, Dr. McKenzie delivered a lecture at Queen's University during which he claimed that "never before have I encountered someone more excited about a work of English Literature than David was about the comet poem by Hopkins." Indeed I was tremendously excited and pleased with this poem, but not just as a work of literature. I sensed that Hopkins knew the sky very well, and that buried within the lines of the poem was a tracing of the path that comet took. I discovered, some two months later, that articles in both the London Times and the Illustrated London News claimed that:

> "... on Monday next it will be situate [sic] about five degrees to the left of the Pleiades, passing thence between the stars Iota in Auriga and Beta in Taurus ..."

I knew the sky well enough at that time to know that Iota (ι) Aurigae was just slightly fainter than Beta (β) Tauri. I had to argue the point with Dr. McKenzie, who thought that an Iota star would be far fainter than a Beta star, but when I convinced him, he was almost as excited as I was and recommended that the comet poem become the high point of my Master's thesis.

Years later, by the time I finally set about completing my Ph.D. dissertation, my astronomical career was in a far better place. I now had 21 comet discoveries to my name, including the co-discovery of Comet Shoemaker-Levy 9, which collided with Jupiter during the summer of 1994. (Another discovery, that of Comet 255P/Levy, would come my way in October 2006, and in 2010 my wife Wendee and I would share in the discovery of yet another, Comet P/2010 E2 Jarnac.) For the new thesis I chose Shakespeare and writers of his time, and how events in the sky inspired some of their writings.

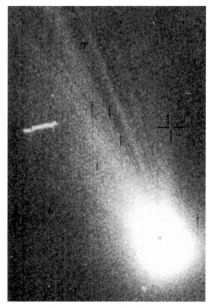

Comet Okazaki-Levy-Rudenko, photographed in 1989 using the Catalina 61-inch telescope atop Mount Bigelow in the Catalina Mountains, north of Tucson, Arizona. (David H. Levy)

A Time for Comets

All of the comets that I have found have been special, like children, to me. The most important find was clearly Comet Shoemaker-Levy 9, a discovery I shared with Eugene and Carolyn Shoemaker and which marked the first time in recorded history that a comet was seen colliding with a planet. This comet was famous not

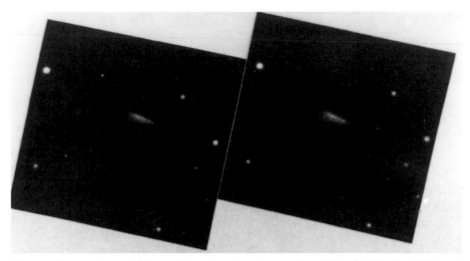

The two discovery images of Comet Shoemaker-Levy 9 captured on 23 March 1993. (David H. Levy)

for what it was but for what it did. In its week-long collision, it completely rewrote our understanding of the interactions among bodies within the solar system, and even of the role of cosmic impacts in the origins of life.

I began searching for comets at a time when visual comets still accounted for a substantial portion of all new comet finds, but as I write these words, visual discoveries are now very rare and as much as I enjoy the search, I would not recommend it for anyone now. Even when I began in 1965, the idea of searching for comets meant more to me than actually discovering a comet. I still have the observing notes I wrote on the night of 17 December that year. The primary goal of the project was not, at the time, to discover a comet, but instead to learn the sky through an orderly search for comets. Actually discovering a new comet was a secondary objective back then.

Although comets have always been my primary interest in astronomy, they are far from the only interest. In fact, variable stars have also played a substantial role in my observing hours, and my search program was called CN3, where C obviously stands for comets, but N means novae. An organized search for novae has never been a part of the program, but I did make independent discoveries of two bright novae. In August 1975 I spotted what appeared to be a slow-moving satellite near Deneb as I prepared to play a game of cards with my then fiancé. (That "practice marriage" did not last very long, by the way.) As I walked indoors, I noted that the satellite was very slow moving. I went out again and saw that it was not moving at all. It turned out to be V1500 Cygni, Nova Cygni 1975. Three years later, with that marriage in ruins, I observed an extra star in the field of the semi-regular variable star W Cygni that turned out to be V1668 Cygni, or Nova Cygni 1978.

Of all the variable stars I have observed or discovered, by far my favourite is TV Corvi, or Tombaugh's Star as I like to call it. That cataclysmic variable star entered my life in 1986. I was writing a biography of my friend Clyde Tombaugh, the discoverer of Pluto, and during interviews I learned that he had discovered a nova in Corvus in 1931. Because Corvus is not a constellation known for novae,

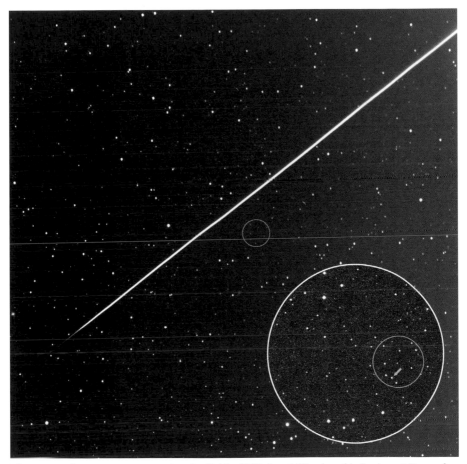

This image shows a meteor crossing the field of TV Corvi (Tombaugh's Star) on 1 December 2005, when the star was at minimum brightness and below the limiting magnitude of the image. The inset shows the star in outburst on 12 June 2004. (David H. Levy)

I wanted to identify it. So in 1987 I visited the plate archives of Lowell Observatory and began a check of the notes Tombaugh took for every one of the photographic plates he took. I found that on 22 March (23 UT) 1931, the great observer had discovered a star undergoing an outburst, within the constellation of Corvus. His plate notes:

> 1 nova suspect. 'T 12' near southwest corner of plate, magnitude about 12 ... No trace of object on 13-inch plates of March 20 and 17, 1931. The image is exactly deformed, like the other star images in the neighborhood. Evidently a very remarkable star to rise from 17 or fainter to 12 in 2 days' time.

On 11 September 1989, I visited the plate stacks at the Harvard College Observatory. In three days of work I examined the position of the nova on more than 260 of their patrol plates. This collection spanned the period from 1930 to 1988, and on nine of those plates I found the star in outburst again and again. In this archival way I determined that the star is a cataclysmic variable star whose outbursts probably could still be visible. Finally, on 23 March 1990, 59 years to the day after Tombaugh first photographed it, I visually detected the star in outburst again, and I was able to get other observers to obtain spectra at that time. The next day, the Central Bureau for Astronomical Telegrams issued Circular 4983, "Cataclysmic Variable in Corvus." Since then I have captured the variable in outburst many times, including once again on a 23 March, and on another night I photographed Tombaugh's Star as just as it was beginning an outburst. By putting together several images from that night I created a movie of the star entering outburst. And as recently as May 2020, I captured the star rolling over in its sleep, as it were, not in outburst but a magnitude brighter than its minimum brightness of about magnitude 19.5. While at minimum, this system is a binary star consisting of a white dwarf and a donor brown dwarf star, orbiting each other in a period of about 90 minutes, and probably less than the diameter of the Sun. The system lies about 1,150 light years away.

I stood and stared ...

Much as I enjoy variable star observing, over the decades I have had to narrow my interests in observing somewhat. TV Corvi, as my favourite variable star, is a star I still follow religiously, along with a few others, like Betelgeuse and T Coronae Borealis. The aurora borealis, or northern lights, have also provided untold hours of joy, from my first display in 1961 to a major aurora including dancing rays, arcs, and spots during the summer of 1966. I even saw one in 2014 from East Jordan, Michigan, on my way to a lecture. During the daytime I follow the Sun, recording

An aurora visible from East Jordan, Michigan on 18 September 2014. (David H. Levy)

occasional sunspots, and a steady appearance of prominences. Eclipses of the Sun and Moon always attract my attention as well. Over many decades I have seen no fewer than 97 eclipses, ranging all the way from tiny barely detectable penumbral eclipses of the Moon to several magnificent total eclipses of the Sun.

One might surmise that nothing can surpass a total eclipse of the Sun. But I have found one thing that can. When I am outside, resting in a lawn chair, I can think of nothing better, nothing more innately peaceful and satisfying, than staring up at the stars, just as the English poet Ralph Hodgson did more than a century ago:

> *I stood and stared, the sky was lit,*
> *The sky was stars all over it,*
> *I stood, I knew not why,*
> *Without a wish, without a will,*
> *I stood upon that silent hill*
> *And stared into the sky until*
> *My eyes were blind with stars, and still*
> *I stared into the sky.*

Internet Satellites
Less Welcome Constellations?

Bob Mizon

Astronomer Royal Sir Martin Rees wrote, for the Commission for Dark Skies' light pollution handbook *Blinded by the Light?* (2020), "The night sky is one part of our environment we have shared with all cultures in all periods of human history. It is very special." Since the dawn of human consciousness, the silent, slowly rotating night sky has inspired in us a sense of wonder, leading us to theorise about what things we cannot engage with might be. We became scientists, poets and musicians partly because we could see the stars, which created within their inner furnaces the elements of which we, and all things, are made. Nowadays that sky is tainted for most of us by light pollution, and more recently by increasing numbers of satellites crossing it, working for us but causing problems not just for astrophotography but for those seeking that calm experience of a pristine starscape.

Our heritage of dark skies: the untainted night sky that has inspired humankind since its beginnings. (Graphic by Kerem Asfuroglu)

There are many kinds of pollution and environmental threats that we have become increasingly concerned about and are increasingly addressing: waste plastics, unwelcome atmospheric gases, deforestation and the biodiversity crash. That half of our environment with so little protection in law, the night sky, also deserves our attention. It must not be that our descendants see stars only on screens and wonder why our generation did not do more to protect them.

On 14 June 2019, Callum Potter, then President of the British Astronomical Association (BAA), the UK's largest body of amateur astronomers, issued a statement[1] expressing the Association's concern at the launching on 23 May 2019 of sixty broadband Internet satellites by the SpaceX company based in Hawthorne, California. The Starlink 'mesh' or 'constellation' is designed to eventually provide worldwide internet coverage at a predicted cost of ten billion dollars. Each satellite of this 'constellation' weighs 227 kilograms (around 500 pounds) and is about the size of a small fridge. Based on a new flat-panel design[2] they use krypton-fuelled plasma thrusters, have high-power antennae, and can autonomously take measures to avoid nearby objects in space.

They are launched in stacks within the payload shroud of the SpaceX Falcon 9 rocket. The satellites orbit at an altitude of about 440 kilometres. Elon Musk, CEO of SpaceX, plans by the mid-2020s to increase the number of Starlink satellites to 12,000, located in three orbital shells: 1,600 at 550 kilometres; around 2,800 at 1,150 kilometres; and around 7,500 at 340 kilometres. The satellites have two symmetrically deployed reflective solar panel arrays. The Starlink passes will interfere with astronomy mostly during the hours around twilight and dawn. Observations that need to be made during the twilight period, such as searches for certain near-Earth asteroids (NEOs), will be compromised. The satellite passes could be visible all night long during the shorter nights of summer.

Other international space and astronomy organisations have expressed their concerns at the potential threat of Starlink to professional and amateur observational astronomy, and warn of the greatly increased number of orbital objects – eventual 'space debris', some say – as SpaceX and other companies with similar ideas, including OneWeb and Amazon, proceed with launches. For example, the International Dark-Sky Association (IDA), the American equivalent of the BAA Commission for Dark Skies, has stated that "the number of low Earth-orbit satellites … has the potential to fundamentally shift the nature of our experience of the night sky… We therefore urge all parties to take precautionary efforts to protect the unaltered night-time environment before deployment of new, large-scale satellite groups." The International Astronomical Union (IAU) has issued a statement detailing "… the problems which large satellite constellations such as Starlink pose for the advance of astronomy." The IAU calls for urgent discussion on

---

1. British Astronomical Association statement: **www.britastro.org/node/18560**
2. Starlink flat-panel design: **spaceflightnow.com/2019/05/15/spacex-releases-new-details-on-starlink-satellite-design**

As Richard Miles photographed Comet C2019 Y4 (Atlas) from darkest Dorset on 23 April 2020, six Starlink satellites passed across the one-quarter of one degree field of view within less than three minutes. (Richard Miles)

ways "… to mitigate or eliminate the detrimental impacts on scientific exploration as soon as practical."

A BAA statement warned that "In the future, if unregulated access continues to be allowed, there may be tens of thousands of spacecraft in low Earth orbit. Already astronomers battle with satellite trails across images and disturbance to visual views of the night sky. With the proposed significant increase of satellites this may become worse by orders of magnitude at certain times of the night. The problem is particularly acute at high latitudes where the satellites are illuminated all night during the summer. The BAA welcomes the unanimity of the astronomical communities, both professional and amateur, who have highlighted these issues."

SpaceX tried at the time of the launch of its first sixty satellites to reassure concerned parties that they would be increasingly less visible from Earth on

The Orion Nebula seen in a pristine night sky from a wood in the Lake District. (Bob Mizon)

reaching their final orbits. Elon Musk tweeted in reply to many questions and criticisms that arose. His messages included:

- 26 May 2019: "If we need to tweak satellite orientation to minimize solar reflection during critical astronomical experiments, that's easily done."
- 27 May 2019: "Sent a note to Starlink team last week specifically regarding albedo reduction. We'll get a better sense of value of this when satellites have raised orbits and arrays are tracking to Sun."
- "Potentially helping billions of economically disadvantaged people is the greater good. That said, we'll make sure Starlink has no material effect on discoveries in astronomy. We care a great deal about science."
- "There are already 4,900 satellites in orbit, which people notice ~0% of the time. Starlink won't be seen by anyone unless looking very carefully and will have ~0% impact on advancements in astronomy."

On 29 April 2020, SpaceX published an "Update on the Starlink Satellite Brightness Issue" explaining to the astronomical community its plans to mitigate the issue of satellite visibility. It stated that: "We firmly believe in the importance of a natural night sky for all of us to enjoy, which is why we have been working with

On the night of 23-24 April 2020 Richard Bassom captured several recently launched Starlink satellites. Brighter trails are satellites in low orbit and the many faint straight lines, about mag +5 in brightness, are likely to be satellites that have reached their final orbits. (Richard Bassom)

leading astronomers around the world to better understand the specifics of their observations and engineering changes we can make to reduce satellite brightness." These include:

- changing the way the satellites fly to their operational altitude, with the satellite knife-edge to the Sun.
- darkening satellites by adding a deployable visor to the satellite to block sunlight from hitting the brightest parts of the spacecraft.

"SpaceX is committed to making future satellite designs as dark as possible. The next generation satellite, designed to take advantage of Starship's unique launch capabilities, will be specifically designed to minimize brightness while also increasing the number of consumers that it can serve with high speed internet access."

Time will allow us to judge whether these promises will be kept, and whether the satellites can be configured to minimise reflections and cause no hindrance to the progress of astronomy.

Meanwhile, the launches continue. On 6 January 2020 SpaceX delivered another sixty Starlink satellites into Earth orbit. One experimental type is called DarkSat. It

A Starlink launch stream imaged from La Palma, Canary Islands on 11 November 2019. (Denis Vida/Global Meteor Network)

is partially painted black. The probe is testing one strategy to reduce the brightness of satellite constellations. On 8 January 2020 the American Astronomical Society held a conference in Honolulu, Hawaii on the subject of internet satellites and their potential impact on astronomy, with special reference to how they might impede programmes undertaken by various telescopes around the world. Jeffrey Hall, Chairman of the Society's Committee on Light Pollution and Director of the Lowell Observatory in Flagstaff, Arizona, stated that: "2020 is the window to figure out what makes a difference in reducing the impact." Patricia Cooper, SpaceX's Vice President in Satellite Government Affairs, leads SpaceX's satellite regulatory, policy and worldwide licensing activities. She told the meeting that "SpaceX is absolutely committed to finding a way forward so our Starlink project doesn't impede the value of the research you all are undertaking."

OneWeb, formerly known as WorldVu Satellites, is a global communications company founded by Greg Wyler, with headquarters in London, England and McLean, Virginia. The company has stated that it intends to be a "thought leader in responsible space" and intends placing its internet satellites at an orbit of 1,200 kilometres, specifically to obviate interference with astronomical observations. OneWeb Vice President Ruth Pritchard-Kelly said: "We chose an orbit as part of our dedication to responsible use of outer space ... and we've also talked to the astronomy community before we launched to make sure that that our satellites won't be too reflective, and that there won't be radio interference with their radio astronomy … it shouldn't be a case of having to choose between connectivity and astronomy."

On the subject of potential space debris proliferation, Stijn Lemmens, Senior Space Debris Mitigation Analyst at the European Space Agency in Darmstadt, has warned that compliance with operational guidelines and norms has not to date always been the case with what he terms 'current actors in spaceflight'; will operators of multi-satellite meshes behave better? Will SpaceX claims to have advanced systems for space debris avoidance prove effective? Musk stated that the Starlink orbiters are "… designed to be capable of fully autonomous collision avoidance – meaning zero humans in the loop."

All information in this article is correct at the time of writing. Updates on the ongoing situation will be available via the Facebook group Action Against Satellite Light Pollution.[3]

Further Information

Information on the continuing debate about satellite constellations and reactions to projects involving them can be found at:

American Astronomical Society Statement: **aas.org/press/aas-issues-position-statement-satellite-constellations**

Association of Universities for Research in Astronomy statement: **www.aura-information astronomy.org/news/aura-statement-on-the-starlink-constellation-of-satellites**

European Southern Observatory statement: **www.eso.org/public/announcements/ann19029**

International Astronomical Union media release: **www.iau.org/news/announcements/detail/ann19035**

International Dark-Sky Association response: **www.darksky.org/starlink-response**

Legacy Survey of Space and Time statement: **www.lsst.org/content/lsst-statement-regarding-increased-deployment-satellite-constellations**

National Radio Astronomy Observatory statement: **public.nrao.edu/news/nrao-statement-commsats**

National Radio Astronomy Observatory and Green Bank Observatory statement: **greenbankobservatory.org/joint-nrao-and-gbo-statement**

Royal Astronomical Society statement: **ras.ac.uk/news-and-press/news/ras-statement-starlink-satellite-constellation**

---

3. **www.facebook.com/groups/ActionAgainstSatelliteLightPollution**

Miscellaneous

Some Interesting Variable Stars

Tracie Heywood

You may have considered taking up variable star observing but how should you choose which stars to observe? There are so many variable stars in the night sky and you don't want to waste your time attempting to follow the "boring" ones. Your choice of stars will, of course, depend on the equipment that you have available, but also needs to be influenced by how much time you can set aside for observing.

This article splits some of the more interesting variable stars into three groups. The group that is most suited to you will be determined by how often you can observe each month and for how long you can observe on a clear night. The light curves included have been constructed from observations stored in the Photometry Database of the British Astronomical Association Variable Star Section. Comparison charts for most of these stars can be found on the BAA Variable Star Section website at **www.britastro.org/vss**

One-Nighters

These are stars that can go through most of their brightness variations in the course of a single (reasonably long) night and would suit *people who can only observe occasionally, but can then observe well into the night.*

STAR	TYPE	RA		DEC		MAX / MIN	PERIOD
		H	M	°	'		
Beta (β) Persei (Algol)	EA	03	08	+40	57	2.1 / 3.4	2.867 days
RS Canum Venaticorum	EA/RS	13	11	+35	56	7.9 / 9.1	4.7977 days
RR Lyrae	RR Lyr	19	26	+42	47	7.0 / 8.1	0.567 days (~13 hours)
HU Tauri	EA	04	38	+20	41	5.9 / 6.7	2.056 days

Beta Persei (Algol) is an eclipsing variable that shows deep primary eclipses but only very shallow secondary eclipses. Although circumpolar from the UK, it is too low in the sky for observation from April to June.

RS Canum Venaticorum is an Algol-type eclipsing variable whose brightness variations can be followed using large binoculars or a small telescope. "Irregularities" can occur throughout the light curve due to the appearance and disappearance of large starspots. Predictions for upcoming eclipses can be found at **www.as.up. krakow.pl/minicalc/CVNRS.HTM**

RR Lyrae is a pulsating variable star that is similar to the classical Cepheid variables. Its variations can however vary slightly from one cycle to the next.

HU Tauri, located close to the Hyades star cluster, is an Algol-type eclipsing variable whose brightness variations can be followed using binoculars. Predictions for upcoming eclipses can be found at **www.as.up.krakow.pl/minicalc/TAUHU. HTM**

Most Clear Nights
These are stars that vary a bit more slowly, but which can display significant changes over a week or two. They would suit *people who can observe for a short while on (nearly) every clear night.*

STAR	TYPE	RA		DEC		MAX / MIN	PERIOD
		H	M	°	'		
R Coronae Borealis	RCB	15	49	+28	09	5.8 / 15.2	None
TV Corvi	UGSU	12	20	−18	47	12.2 / 19.25	90.55 minutes (orbital)
Zeta (ζ) Geminorum	Cepheid	07	04	+20	34	3.7 / 4.2	10.1498 days
AG Pegasi	Z And	21	51	+12	38	6.0 / 9.4	816 days
GK Persei	U Gem	03	31	+43	54	9.7 / 14.0	1.997 days (orbital)

R Coronae Borealis spends most of its time close to magnitude 6.0, but at times unpredictable in advance it shows dramatic fades, sometimes going all the way down to 15th magnitude. Eventually it starts to brighten again and returns to maximum. A fade which started in 2007 was unusually prolonged, lasting for more than a decade.

TV Corvi is a dwarf nova that was discovered by Clyde Tombaugh in 1932. Normally around 19th magnitude, the brightest outbursts have reached 12th magnitude.

Zeta (ζ) Geminorum is a classic Cepheid variable. As with other classic Cepheids the brightness changes repeat exactly from one 10.1498 day cycle to the next.

AG Pegasi spends most of its time close to magnitude 9.0. A series of outbursts during the second half of the nineteenth century eventually raised it to magnitude 6.0 before a slow fade set in. More recently, it brightened to around magnitude 7.0 during the summer of 2015. The 816-day period listed is an orbital period during which the brightness fluctuates by several tenths of a magnitude.

GK Persei produced a classical nova outburst in 1901 before returning to minimum at 13th magnitude. In recent decades it has produced smaller dwarf nova outbursts every few years. Recent outbursts have occurred in spring 2010, spring 2015 and autumn 2018.

Several Times per Month
Slower variables whose brightness will change significantly over several months or a year. These variables would suit *people who can observe several times per month, but not necessarily on every clear night.*

STAR	TYPE	RA		DEC		MAX / MIN	PERIOD
		H	M	°	'		
RS Cancri	SR	09	11	+30	58	4.8 / 6.9	242 days
RY Draconis	SR	12	56	+66	00	5.9 / 8.0	~300 days
TV Geminorum	Irregular	06	12	+21	52	6.3 / 7.8	None
AC Herculis	RVA	18	30	+21	52	6.9 / 9.0	75.29 days
R Serpentis	Mira	15	51	+15	08	5.2 / 14.4	356 days
R Trianguli	Mira	02	37	+34	16	5.4 / 12.6	267 days
SS Virginis	SR	12	25	+00	46	6.0 / 9.6	~360 days

RS Cancri and **RY Draconis** are semi-regular variables. "Semi-regular" indicates that the brightness variations only roughly repeat from one cycle to the next. The listed period is only an average and the observed brightness range will differ between cycles, sometimes covering the whole of the listed range, but usually being somewhat smaller. Long-term studies often reveal the presence of additional periods.

TV Geminorum is listed as being an irregular variable, but some more recent studies have suggested the presence of a period close to 8 years as well as shorter periods of around 8 months and 14 months.

AC Herculis is a RV Tauri type variable. These are pulsating stars that typically show alternate shallow and deep minima over a period of several months. Not all deep minima are equally deep – a one magnitude drop is the most common.

R Serpentis and **R Trianguli** are Mira type variables. R Serpentis has an average period of 356 days and will start 2022 near minimum. It is due to reach maximum during May when Serpens is well placed in the evening sky. R Trianguli will have two maxima during 2022. The first will be during January, with the second occurring in the early autumn.

SS Virginis used to be listed as a Mira type variable with a period very close to a year, although more recently, it has been realized that it is a semi-regular variable with a period of around 360 days. The 2022 peak is likely to occur during early March with Virgo quite well placed in the evening sky.

Light curve for RR Lyrae in 2015

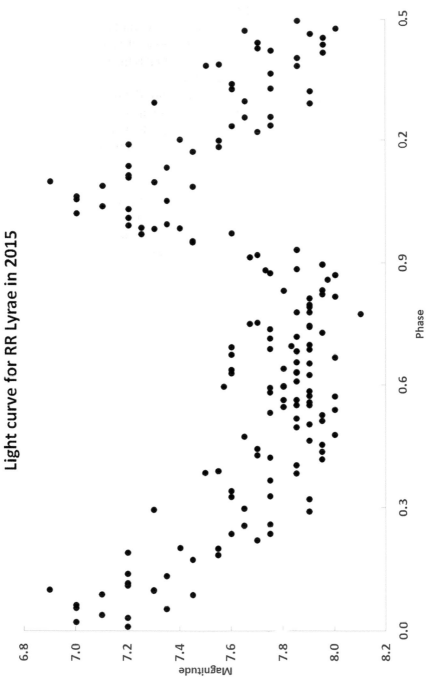

The author's light curve for RR Lyrae during 2015. All observations made during the year have been combined into this single light curve showing 1.5 cycles of variation. (Tracie Heywood)

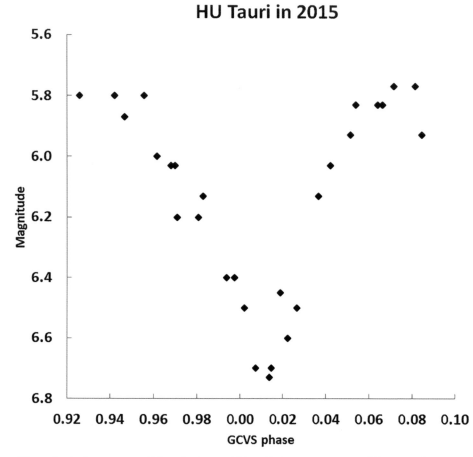

The author's observations of the eclipsing variable HU Tauri in 2015 around the time of primary eclipse. (Tracie Heywood)

Light Curve for GK PER

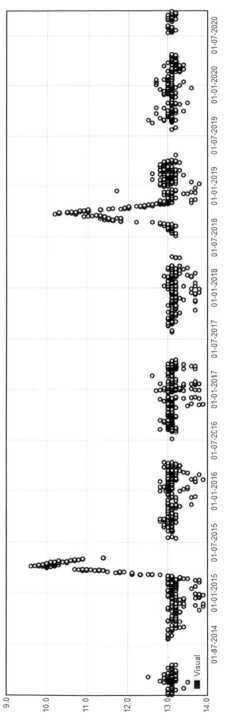

The BAA VSS light curve for the dwarf nova GK Persei from 2014 to 2020. (BAA Photometry Database)

Light Curve for RY DRA

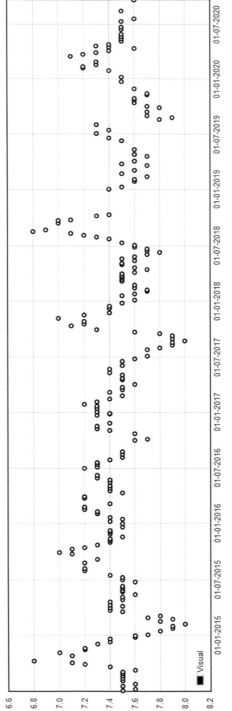

The author's light curve for the semi-regular variable RY Draconis from summer 2014 to summer 2020. (Tracie Heywood)

Light Curve for TV GEM

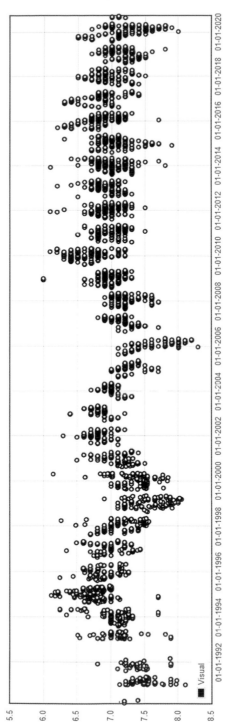

A 30-year BAA VSS light curve for the semi-regular variable TV Geminorum. (BAA Photometry Database)

Light Curve for R SER

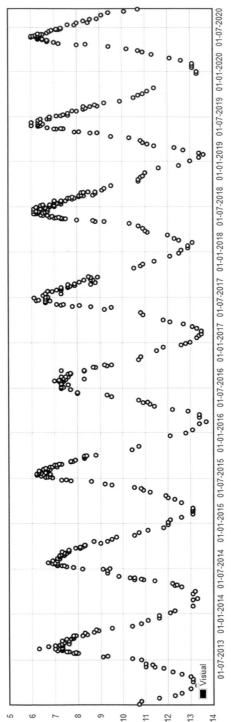

The BAA VSS light curve for the Mira type variable R Serpentis from 2013 to 2020. (BAA Photometry Database)

R TRIANGULI

02h 37m 02.3s +34° 15' 51" (2000)

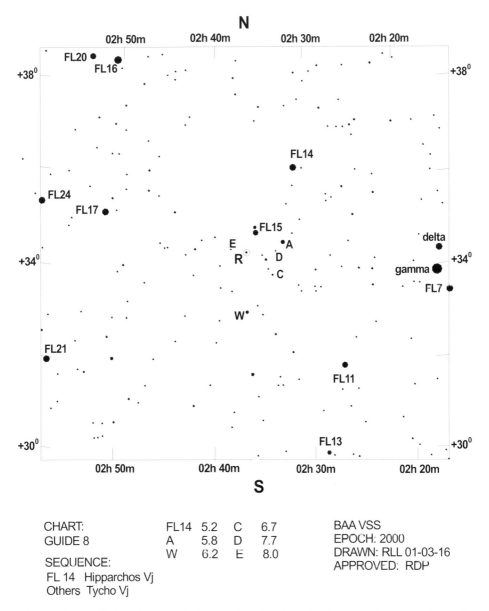

CHART:	FL14	5.2	C	6.7	BAA VSS
GUIDE 8	A	5.8	D	7.7	EPOCH: 2000
	W	6.2	E	8.0	DRAWN: RLL 01-03-16
SEQUENCE:					APPROVED: RDP
FL 14 Hipparchos Vj					
Others Tycho Vj					

The BAA VSS finder chart for R Trianguli, with six suitable comparison stars labelled. (BAA Variable Star Section)

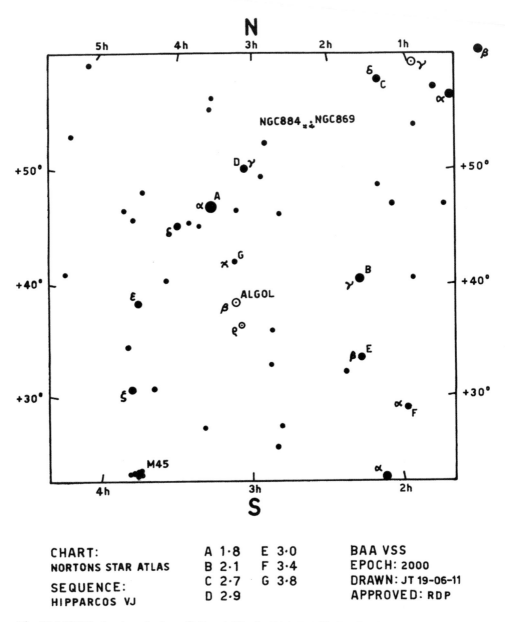

The BAA VSS finder chart for Beta (β) Persei (Algol). (BAA Variable Star Section)

Minima of Algol in 2022

Beta (β) Persei (Algol): Magnitude 2.1 to 3.4 / Duration 9.6 hours

		h				h				h				h	
Jan	3	0.1	★	Feb	3	13.1		Mar	1	8.4		Apr	1	21.4	★
	5	20.9	★		6	9.9			4	5.2			4	18.2	
	8	17.7	★		9	6.7			7	2.0			7	15.0	
	11	14.5			12	3.5			9	22.8	★		10	11.8	
	14	11.3			15	0.3	★		12	19.7	★		13	8.6	
	17	8.2			17	21.1	★		15	16.5			16	5.5	
	20	5.0			20	17.9			18	13.3			19	2.3	
	23	1.8	★		23	14.8			21	10.1			21	23.1	
	25	22.6	★		26	11.6			24	6.9			24	19.9	
	28	19.4	★						27	3.7			27	16.7	
	31	16.2							30	0.6			30	13.5	
May	3	10.4		Jun	1	2.5		Jul	2	15.5		Aug	3	4.5	
	6	7.2			3	23.3			5	12.3			6	1.3	★
	9	4.0			6	20.1			8	9.1			8	22.1	
	12	0.8			9	17.0			11	5.9			11	18.9	
	14	21.6			12	13.8			14	2.8			14	15.7	
	17	18.4			15	10.6			16	23.6			17	12.5	
	20	15.2			18	7.4			19	20.4			20	9.4	
	23	12.1			21	4.2			22	17.2			23	6.2	
	26	8.9			24	1.0			25	14.0			26	3.0	
	29	5.7			26	21.9			28	10.8			28	23.8	★
					29	18.7			31	7.7			31	20.6	
Sep	3	17.4		Oct	2	9.6		Nov	2	22.6	★	Dec	1	14.7	
	6	14.3			5	6.4			5	19.4	★		4	11.6	
	9	11.1			8	3.2	★		8	16.2			7	8.4	
	12	7.9			11	0.1	★		11	13.0			10	5.2	
	15	4.7			13	20.9	★		14	9.8			13	2.0	★
	18	1.5	★		16	17.7			17	6.7			15	22.8	★
	20	22.3	★		19	14.5			20	3.5	★		18	19.6	★
	23	19.2			22	11.3			23	0.3	★		21	16.5	
	26	16.0			25	8.1			25	21.1	★		24	13.3	
	29	12.8			28	5.0			28	17.9			27	10.1	
					31	1.8	★						30	6.9	

Eclipses marked with an asterisk (★) are favourable from the British Isles, taking into account the altitude of the variable and the distance of the Sun below the horizon (based on longitude 0° and latitude 52°N)

All times given in the above table are expressed in UT/GMT.

Some Interesting Double Stars

Brian Jones

The accompanying table describes the visual appearances of a selection of double stars. These may be optical doubles (which consist of two stars which happen to lie more or less in the same line of sight as seen from Earth and which therefore only appear to lie close to each other) or binary systems (which are made up of two stars which are gravitationally linked and which orbit their common centre of mass).

Other than the location on the celestial sphere and the magnitudes of the individual components, the list gives two other values for each of the double stars listed – the angular separation and position angle (PA). Further details of what these terms mean can be found in the article *Double and Multiple Stars* published in the 2018 edition of the Yearbook of Astronomy.

Double-star observing can be a very rewarding process, and even a small telescope will show most, if not all, the best doubles in the sky. You can enjoy looking at double stars simply for their beauty, such as Albireo (β Cygni) or Almach (γ Andromedae), although there is a challenge to be had in splitting very difficult (close) double stars, such as the demanding Sirius (α Canis Majoris) or the individual pairs forming the Epsilon (ε) Lyrae 'Double-Double' star system.

The accompanying list is a compilation of some of the prettiest double (and multiple) stars scattered across both the Northern and Southern heavens. Once you have managed to track these down, many others are out there awaiting your attention …

Star	RA		Declination		Magnitudes	Separation	PA	Comments
	h	m	°	'		(arcsec)	°	
Beta[1,2] ($\beta^{1,2}$) Tucanae	00	31.5	−62	58	4.36 / 4.53	27.1	169	Both stars again double, but difficult
Achird (η Cassiopeiae)	00	49.1	+57	49	3.44 / 7.51	13.4	324	Easy double
Mesarthim (γ Arietis)	01	53.5	+19	18	4.58 / 4.64	7.6	1	Easy pair of white stars
Almach (γ Andromedae)	02	03.9	+42	20	2.26 / 4.84	9.6	63	Yellow and blue-green components
32 Eridani	03	54.3	−02	57	4.8 / 6.1	6.9	348	Yellowish and bluish
Alnitak (ζ Orionis)	05	40.7	−01	57	2.0 / 4.3	2.3	167	Difficult, can be resolved in 10cm telescopes
Gamma (γ) Leporis	05	44.5	−22	27	3.59 / 6.28	95.0	350	White and yellow-orange components, easy pair
Sirius (α Canis Majoris)	06	45.1	−16	43	−1.4 / 8.5			Binary, period 50 years, difficult
Castor (α Geminorum)	07	34.5	+31	53	1.93 / 2.97	7.0	55	Binary, 445 years, widening
Gamma (γ) Velorum	08	09.5	−47	20	1.83 / 4.27	41.2	220	Pretty pair in nice field of stars
Upsilon (υ) Carinae	09	47.1	−65	04	3.08 / 6.10	5.03	129	Nice object in small telescopes
Algieba (γ Leonis)	10	20.0	+19	50	2.28 / 3.51	4.6	126	Binary, 510 years, orange-red and yellow
Acrux (α Crucis)	12	26.4	−63	06	1.40 / 1.90	4.0	114	Glorious pair, third star visible in low power
Porrima (γ Virginis)	12	41.5	−01	27	3.56 / 3.65			Binary, 170 years, widening, visible in small telescopes
Cor Caroli (α Canum Venaticorum)	12	56.0	+38	19	2.90 / 5.60	19.6	229	Easy, yellow and bluish
Mizar (ζ Ursae Majoris)	13	24.0	+54	56	2.3 / 4.0	14.4	152	Easy, wide naked-eye pair with Alcor
Alpha (α) Centauri	14	39.6	−60	50	0.0 / 1.2			Binary, beautiful pair of stars
Izar (ε Boötis)	14	45.0	+27	04	2.4 / 5.1	2.9	344	Fine pair of yellow and blue stars
Omega[1,2] ($\omega^{1,2}$) Scorpii	16	06.0	−20	41	4.0 / 4.3	14.6	145	Optical pair, easy
Epsilon[1] (ε^1) Lyrae	18	44.3	+39	40	4.7 / 6.2	2.6	346	The Double-Double, quadruple system with ε^2
Epsilon[2] (ε^2) Lyrae	18	44.3	+39	40	5.1 / 5.5	2.3	76	Both individual pairs just visible in 80mm telescopes
Theta[1,2] ($\theta^{1,2}$) Serpentis	18	56.2	+04	12	4.6 / 5.0	22.4	104	Easy pair, mag 6.7 yellow star 7 arc minutes from θ^2

Star	RA		Declination		Magnitudes	Separation	PA	Comments
	h	m	°	′		(arcsec)	°	
Albireo (β Cygni)	19	30.7	+27	58	3.1 / 5.1	34.3	54	Glorious pair, yellow and blue-green
Algedi (α¹·² Capricorni)	20	18.0	−12	32	3.7 / 4.3	6.3	292	Optical pair, easy
Gamma (γ) Delphini	20	46.7	+16	07	5.14 / 4.27	9.2	265	Easy, orange and yellow-white
61 Cygni	21	06.9	+38	45	5.20 / 6.05	31.6	152	Binary, 678 years, both orange
Theta (θ) Indi	21	19.9	−53	27	4.6 / 7.2	7.0	275	Fine object for small telescopes
Delta (δ) Tucanae	22	27.3	−64	58	4.49 / 8.7	7.0	281	Beautiful double, white and reddish

Some Interesting Nebulae, Star Clusters and Galaxies

Brian Jones

Object	RA h	RA m	Declination °	Declination '	Remarks
47 Tucanae (in Tucana)	00	24.1	−72	05	Fine globular cluster, easy with naked eye
M31 (in Andromeda)	00	40.7	+41	05	Andromeda Galaxy, visible to unaided eye
Small Magellanic Cloud	00	52.6	−72	49	Satellite galaxy of the Milky Way
NGC 362 (in Tucana)	01	03.3	−70	51	Globular cluster, impressive sight in telescopes
M33 (in Triangulum)	01	31.8	+30	28	Triangulum Spiral Galaxy, quite faint
NGC 869 and NGC 884	02	20.0	+57	08	Sword Handle Double Cluster in Perseus
M34 (in Perseus)	02	42.1	+42	46	Open star cluster near Algol
M45 (in Taurus)	03	47.4	+24	07	Pleiades or Seven Sisters cluster, a fine object
Large Magellanic Cloud	05	23.5	−69	45	Satellite galaxy of the Milky Way
30 Doradus (in Dorado)	05	38.6	−69	06	Star-forming region in Large Magellanic Cloud
M1 (in Taurus)	05	32.3	+22	00	Crab Nebula, near Zeta (ζ) Tauri
M38 (in Auriga)	05	28.6	+35	51	Open star cluster
M42 (in Orion)	05	33.4	−05	24	Orion Nebula
M36 (in Auriga)	05	36.2	+34	08	Open star cluster
M37 (in Auriga)	05	52.3	+32	33	Open star cluster
M35 (in Gemini)	06	06.5	+24	21	Open star cluster near Eta (η) Geminorum
M41 (in Canis Major)	06	46.0	−20	46	Open star cluster to south of Sirius
M44 (in Cancer)	08	38.0	+20	07	Praesepe, visible to naked eye
M81 (in Ursa Major)	09	55.5	+69	04	Bode's Galaxy
M82 (in Ursa Major)	09	55.9	+69	41	Cigar Galaxy or Starburst Galaxy
Carina Nebula (in Carina)	10	45.2	−59	52	NGC 3372, large area of bright and dark nebulosity
M104 (in Virgo)	12	40.0	−11	37	Sombrero Hat Galaxy to south of Porrima
Coal Sack (in Crux)	12	50.0	−62	30	Prominent dark nebula, visible to naked eye
NGC 4755 (in Crux)	12	53.6	−60	22	Jewel Box open cluster, magnificent object
Omega (ω) Centauri	13	23.7	−47	03	Splendid globular in Centaurus, easy with naked eye
M51 (in Canes Venatici)	13	29.9	+47	12	Whirlpool Galaxy
M3 (in Canes Venatici)	13	40.6	+28	34	Bright Globular Cluster

Object	RA		Declination		Remarks
	h	m	°	'	
M4 (in Scorpius)	16	21.5	−26	26	Globular cluster, close to Antares
M12 (in Ophiuchus)	16	47.2	−01	57	Globular cluster
M10 (in Ophiuchus)	16	57.1	−04	06	Globular cluster
M13 (in Hercules)	16	40.0	+36	31	Great Globular Cluster, just visible to naked eye
M92 (in Hercules)	17	16.1	+43	11	Globular cluster
M6 (in Scorpius)	17	36.8	−32	11	Open cluster
M7 (in Scorpius)	17	50.6	−34	48	Bright open cluster
M20 (in Sagittarius)	18	02.3	−23	02	Trifid Nebula
M8 (in Sagittarius)	18	03.6	−24	23	Lagoon Nebula, just visible to naked eye
M16 (in Serpens)	18	18.8	−13	49	Eagle Nebula and star cluster
M17 (in Sagittarius)	18	20.2	−16	11	Omega Nebula
M11 (in Scutum)	18	49.0	−06	19	Wild Duck open star cluster
M57 (in Lyra)	18	52.6	+32	59	Ring Nebula, brightest planetary
M27 (in Vulpecula)	19	58.1	+22	37	Dumbbell Nebula
M29 (in Cygnus)	20	23.9	+38	31	Open cluster
M15 (in Pegasus)	21	28.3	+12	10	Bright globular cluster near Epsilon (ε) Pegasi
M39 (in Cygnus)	21	31.6	+48	25	Open cluster, good with low powers
M52 (in Cassiopeia)	23	24.2	+61	35	Open star cluster near 4 Cassiopeiae

M = Messier Catalogue Number   NGC = New General Catalogue Number

The positions in the sky of each of the objects contained in this list are given on the Monthly Star Charts printed elsewhere in this volume.

Astronomical Organizations

American Association of Variable Star Observers

49 Bay State Road, Cambridge, Massachusetts 02138, USA

www.aavso.org

The AAVSO is an international non-profit organization of variable star observers whose mission is to enable anyone, anywhere, to participate in scientific discovery through variable star astronomy. We accomplish our mission by carrying out the following activities:

- observation and analysis of variable stars
- collecting and archiving observations for worldwide access
- forging strong collaborations between amateur and professional astronomers
- promoting scientific research, education and public outreach using variable star data

American Astronomical Society

1667 K Street NW, Suite 800, Washington, DC 20006, USA

https://aas.org

Established in 1899, the American Astronomical Society (AAS) is the major organization of professional astronomers in North America. The mission of the AAS is to enhance and share humanity's scientific understanding of the universe, which it achieves through publishing, meeting organization, education and outreach, and training and professional development.

Astronomical Society of the Pacific

390 Ashton Avenue, San Francisco, CA 94112, USA

www.astrosociety.org

Formed in 1889, the Astronomical Society of the Pacific (ASP) is a non-profit membership organization which is international in scope. The mission of the ASP is to increase the understanding and appreciation of astronomy through the engagement of our many constituencies to advance science and science literacy. We invite you to explore our site to learn more about us; to check out our resources

336 Yearbook of Astronomy 2022

and education section for the researcher, the educator, and the backyard enthusiast; to get involved by becoming an ASP member; and to consider supporting our work for the benefit of a science literate world!

Astrospeakers.org
www.astrospeakers.org

A website designed to help astronomical societies and clubs locate astronomy and space lecturers which is also designed to help people find their local astronomical society. It is completely free to register and use and, with over 50 speakers listed, is an excellent place to find lecturers for your astronomical society meetings and events. Speakers and astronomical societies are encouraged to use the online registration to be added to the lists.

British Astronomical Association
Burlington House, Piccadilly, London, W1J 0DU, England
www.britastro.org

The British Astronomical Association is the UK's leading society for amateur astronomers catering for beginners to the most advanced observers who produce scientifically useful observations. Our Observing Sections provide encouragement and advice about observing. We hold meetings around the country and publish a bi-monthly Journal plus an annual Handbook. For more details, including how to join the BAA or to contact us, please visit our website.

British Interplanetary Society
Arthur C Clarke House, 27/29 South Lambeth Road, London, SW8 1SZ, England
www.bis-space.com

The British Interplanetary Society is the world's longest-established space advocacy organisation, founded in 1933 by the pioneers of British astronautics. It is the first organisation in the world still in existence to design spaceships. Early members included Sir Arthur C Clarke and Sir Patrick Moore. The Society has created many original concepts, from a 1938 lunar lander and space suit designs, to geostationary orbits, space stations and the first engineering study of a starship, Project Daedalus. Today the BIS has a worldwide membership and welcomes all with an interest in Space, including enthusiasts, students, academics and professionals.

Canadian Astronomical Society
Société Canadienne D'astronomie (CASCA)
100 Viaduct Avenue West, Victoria, British Columbia, V9E 1J3, Canada
www.casca.ca
CASCA is the national organization of professional astronomers in Canada. It seeks to promote and advance knowledge of the universe through research and education. Founded in 1979, members include university professors, observatory scientists, postdoctoral fellows, graduate students, instrumentalists, and public outreach specialists.

Royal Astronomical Society of Canada
203-4920 Dundas St W, Etobicoke, Toronto, ON M9A 1B7, Canada
www.rasc.ca
Bringing together over 5,000 enthusiastic amateurs, educators and professionals RASC is a national, non-profit, charitable organization devoted to the advancement of astronomy and related sciences and is Canada's leading astronomy organization. Membership is open to everyone with an interest in astronomy. You may join through any one of our 29 RASC centres, located across Canada and all of which offer local programs. The majority of our events are free and open to the public.

Federation of Astronomical Societies
The Secretary, 147 Queen Street, SWINTON, Mexborough, S64 8NG
www.fedastro.org.uk
The Federation of Astronomical Societies (FAS) is an umbrella group for astronomical societies in the UK. It promotes cooperation, knowledge and information sharing and encourages best practice. The FAS aims to be a body of societies united in their attempts to help each other find the best ways of working for their common cause of creating a fully successful astronomical society. In this way it endeavours to be a true federation, rather than some remote central organization disseminating information only from its own limited experience. The FAS also provides a competitive Public Liability Insurance scheme for its members.

International Dark-Sky Association
darksky.org
The International Dark-Sky Association (IDA) is the recognized authority on light pollution and the leading organization combating light pollution worldwide. The IDA works to protect the night skies for present and future generations, our public outreach efforts providing solutions, quality education and programs that inform

audiences across the United States of America and throughout the world. At the local level, our mission is furthered through the work of our U.S. and international chapters representing five continents.

The goals of the IDA are:

- Advocate for the protection of the night sky
- Educate the public and policymakers about night sky conservation
- Promote environmentally responsible outdoor lighting
- Empower the public with the tools and resources to help bring back the night

Royal Astronomical Society of New Zealand

PO Box 3181, Wellington, New Zealand

www.rasnz.org.nz

Founded in 1920, the object of The Royal Astronomical Society of New Zealand is the promotion and extension of knowledge of astronomy and related branches of science. It encourages interest in astronomy and is an association of observers and others for mutual help and advancement of science. Membership is open to all interested in astronomy. The RASNZ has about 180 individual members including both professional and amateur astronomers and many of the astronomical research and observing programmes carried out in New Zealand involve collaboration between the two. In addition the society has a number of groups or sections which cater for people who have interests in particular areas of astronomy.

Astronomical Society of Southern Africa

Astronomical Society of Southern Africa, c/o SAAO, PO Box 9, Observatory, 7935, South Africa

assa.saao.ac.za

Formed in 1922, The Astronomical Society of Southern Africa comprises both amateur and professional astronomers. Membership is open to all interested persons. Regional Centres host regular meetings and conduct public outreach events, whilst national Sections coordinate special interest groups and observing programmes. The Society administers two Scholarships, and hosts occasional Symposia where papers are presented. For more details, or to contact us, please visit our website.

Royal Astronomical Society

Burlington House, Piccadilly, London, W1J 0BQ, England

www.ras.org.uk

The Royal Astronomical Society, with around 4,000 members, is the leading UK body representing astronomy, space science and geophysics, with a membership including professional researchers, advanced amateur astronomers, historians of science, teachers, science writers, public engagement specialists and others.

Society for Popular Astronomy

Secretary: Guy Fennimore, 36 Fairway, Keyworth, Nottingham, NG12 5DU

www.popastro.com

The Society for Popular Astronomy is a national society that aims to present astronomy in a less technical manner. The bi-monthly society magazine Popular Astronomy is issued free to all members.

Our Contributors

Martin Braddock is a professional scientist and project leader in the field of drug discovery and development with 36 years' experience of working in academic institutes and large corporate organizations. He holds a BSc in Biochemistry and a PhD in Radiation Biology and is a former Royal Society University Research Fellow at the University of Oxford. He was elected a Fellow of the Royal Society of Biology in 2010, and in 2012 received an Alumnus Achievement Award for distinction in science from the University of Salford. Martin has published over 190 peer-reviewed scientific publications, filed nine patents, and edited two books for the Royal Society of Chemistry and serves as a grant proposal evaluator for multiple international research agencies. Martin holds further qualifications from the University of Central Lancashire and Open University. He is a member of the Mansfield and Sutton Astronomical Society and was elected a Fellow of the Royal Astronomical Society in May 2015 and has given over 100 lectures to astronomical societies in the U.K. An ambassador for science, technology, engineering and mathematics (STEM), Martin seeks to inspire the next generation of young scientists to aim high and be the best they can be. To find out more about him visit **science4u.co.uk**

Steve Brown is an amateur astronomer based in Stokesley in the North Riding of Yorkshire. He has been interested in astronomy from a young age and has observed seriously since buying his first telescope (a 130mm reflector) in 2011. He regularly observes from his garden (when the Yorkshire weather allows) dividing his observing time between astrophotography and sketching. His image 'The Rainbow Star' won the Stars and Nebulae category of the Insight Astronomy Photographer of the Year competition in 2016. Steve also appeared on the Sky at Night episode 'Moore Winter Marathon', filmed at Kielder Observatory and broadcast in March 2013. The programme also featured several of his astro sketches. Steve has also had sketches, images and articles published in *Sky at Night* and *All About Space* magazines, as well as in a number of astronomy books. As a member of the British Astronomical Association, Steve regularly submits meteor shower observations to their Meteor Section. You can follow Steve on Twitter via **@sjb_astro** and see his images on Flickr at **www.flickr.com/photos/sjb_astro**

Allan Chapman is a historian of science at Oxford University, with a special interest in the history of astronomy. He has written many books and academic articles, lectures widely, and has made several television programmes. In 2015 he was presented with the Jackson-Gwilt Medal by the Royal Astronomical Society. His books include *The Victorian Amateur Astronomer: Independent Astronomical Research in Britain 1820–1920* (1998, 2017), *Stargazers: Copernicus, Galileo, the Telescope, and the Church. The Astronomical Renaissance, 1500–1700* (2014), and *Comets, Cosmology, and the Big Bang: A History of Astronomy from Edmond Halley to Edwin Hubble, 1700–2000* (2018).

Neil Haggath has a degree in astrophysics from Leeds University and has been a Fellow of the Royal Astronomical Society since 1993. A member of Cleveland and Darlington Astronomical Society for 41 years, he has served on its committee for 33 years. Neil is an avid umbraphile, clocking up six total eclipse expeditions so far, to locations as far flung as Australia and Hawai'i. Four of them were successful, the most recent being in Jackson, Wyoming on 21 August 2017. In 2012, he may have set a somewhat unenviable record among British astronomers – for the greatest distance travelled (6,000 miles to Thailand) to NOT see the transit of Venus. He saw nothing on the day … and got very wet!

Dr. David M. Harland gained his BSc in astronomy in 1977, lectured in computer science, worked in industry, and managed academic research. In 1995 he 'retired' in order to write on space themes.

David Harper, FRAS has had a varied career which includes teaching mathematics, astronomy and computing at Queen Mary University of London, astronomical software development at the Royal Greenwich Observatory, bioinformatics support at the Wellcome Trust Sanger Institute, and a research interest in the dynamics of planetary satellites, which began during his Ph.D. at Liverpool University in the 1980s and continues in' an occasional collaboration with colleagues in China. He is married to fellow contributor Lynne Stockman.

Tracie Heywood is an amateur astronomer from Leek in Staffordshire and is one of the UK's leading variable star observers, using binoculars to monitor the brightness changes of several hundred variable stars. Tracie currently writes a monthly column about variable stars for *Astronomy Now* magazine. She has previously been the Eclipsing Binary coordinator for the Variable Star Section of the British Astronomical Association and the Director of the Variable Star Section of the Society for Popular Astronomy.

Richard "Rik" Hill made his first astronomical observation on 5 May 1957 when he watched the transit of Mercury. He observed throughout the 1960s, first with a 2.4″ refractor and later with an RV-6, 6″ f/9 Newtonian reflector. Rik has been a member of both the Association of Lunar & Planetary Observers (ALPO) and American Association of Variable Star Observers (AAVSO) since 1975, and a member of the BAA since the 1980s. In 1982 Walter Haas, Director of the ALPO, asked him to found the Solar Section, which he did, and which he still runs. In 1979, Rik was hired by Warner & Swasey Observatory (Case Western Reserve University) to operate their Burrell Schmidt telescope at Kitt Peak, where he worked for 12 years until the grant was terminated. Then in 1992 he began work with the Planetary Atmospheres and Planetary Occultations groups at the Lunar & Planetary Laboratory (LPL) of the University of Arizona, migrating in 1999 to the Near Earth Asteroid search team Catalina Sky Survey (CSS) also at LPL. He retired in October 2015.

Rod Hine was aged around ten when he was given a copy of *The Boys Book of Space* by Patrick Moore. Already interested in anything to do with science and engineering he devoured the book from cover to cover. The launch of Sputnik I shortly afterwards clinched his interest in physics and space travel. He took physics, chemistry and mathematics at A-level and then studied Natural Sciences at Churchill College, Cambridge. He later switched to Electrical Sciences and subsequently joined Marconi at Chelmsford working on satellite communications in the UK, Middle East and Africa. This led to work in meteorological communications in Nairobi, Kenya and later a teaching post at the Kenya Polytechnic. There he met and married a Yorkshire lass and moved back to the UK in 1976. Since then he has had a variety of jobs in electronics and industrial controls, and until recently was lecturing part-time at the University of Bradford. Rod got fully back into astronomy in around 1992 when his wife bought him an astronomy book, at which time he joined Bradford Astronomical Society. With the call sign G8AQH, he is also a member of Otley Amateur Radio Society. He is currently working part-time at Leeds University providing engineering support for a project to convert redundant satellite dishes into radio telescopes in developing countries.

Brian Jones hails from Bradford in the West Riding of Yorkshire and was a founder member of the Bradford Astronomical Society. He developed a fascination with astronomy at the age of five when he first saw the stars through a pair of binoculars, although he spent the first part of his working life developing a career in mechanical engineering. However, his true passion lay in the stars and his interest in astronomy took him into the realms of writing sky guides for local newspapers, appearing on local radio and television, teaching astronomy and space in schools and, in 1985, leaving engineering to become a full time astronomy and space writer. His books have covered a range of astronomy and space-related topics for both children and adults and his journalistic work includes writing articles and book reviews for several astronomy magazines as well as for many general interest magazines, newspapers and periodicals. His passion for bringing an appreciation of the universe to his readers is reflected in his writing.

You can follow Brian on Twitter via @StarsBrian and check out the sky by visiting his blog at starlight-nights.co.uk from where you can also access his Facebook group Starlight Nights.

Carolyn Kennett lives in the far south-west of Cornwall. She likes to write, although you will often find her walking the Tinner's Way and coastal pathways found in the local countryside. As well as researching local astronomy history, she has a passion for archaeoastronomy and eighteenth- and nineteenth-century astronomy.

David H. Levy is arguably one of the most enthusiastic and best known amateur astronomers of our time. Although he has never taken a class in astronomy, he has received five honorary doctorates in Science and a PhD combining astronomy and English Literature. David has penned over three dozen books and written for three astronomy magazines, as well as appearing on television programmes featured on both the Discovery Channel and Science Channel. Among his other accomplishments are 23 comet discoveries (the most famous being Shoemaker-Levy 9 that collided with Jupiter in 1994); several hundred shared asteroid discoveries; and receiving an Emmy for the documentary *Three Minutes to Impact*. David is the former editor of the web magazine *Sky's Up* and continues to hunt for comets and asteroids.

John McCue graduated in astronomy from the University of St Andrews and began teaching. He gained a Ph.D. from Teesside University studying the unusual rotation of Venus. In 1979 he and his colleague John Nichol founded the Cleveland and Darlington Astronomical Society, which then worked in partnership with the local authority to build the Wynyard Planetarium and Observatory in Stockton-on-Tees. John is currently double star advisor for the British Astronomical Association.

Mary McIntyre is an amateur astronomer, astrophotographer and astronomy sketcher based in Oxfordshire, UK. She has had a life-long interest in astronomy and loves sharing that with others. Mary is a writer and communicator and has contributed several articles to *Sky at Night* magazine and made numerous radio appearances including being a co-presenter of the monthly radio show Comet Watch and a member of the AstroRadio team. She also gives regular astronomy and photography talks to people of all ages, including presentations to astronomy societies, camera clubs, schools and Beavers/Cubs/Scouts groups. She is passionate about promoting science, technology, engineering and mathematics (STEM) and astronomy to women and girls and runs the UK Women in Astronomy Network, which connects, celebrates and promotes women with a passion for astronomy. By providing role models from both professional and the keen amateur, women and young girls are encouraged and supported to explore all aspects of astronomy and astrophotography and given the confidence to pursue a career in the field. To find out more about Mary visit **www.marymcintyreastronomy.co.uk**

Bob Mizon graduated from King's College London in 1969 with a degree in Modern Languages. His abiding interest in astronomy was sparked by a chance encounter at the age of ten with the *Larousse Encyclopaedia of Astronomy* in a primary school library. This culminated in his switching in 1996 from teaching French to operating a mobile schools' planetarium and speaking to groups and societies all over the UK. Bob is a Fellow of the Royal Astronomical Society and was awarded an MBE in 2010 for his work as an astronomy educator and as coordinator of the British Astronomical Association's Commission for Dark Skies (CfDS) which can be seen at **www.britastro.org/dark-skies**. A Lifetime Member of the International Dark-Sky Association (IDA), Bob has worked with CfDS colleagues in support of several of Britain's IDA Dark Sky Reserves and Parks.

Neil Norman, FRAS first became fascinated with the night sky when he was five years of age and saw Patrick Moore on the television for the first time. It was the Sky at Night programme, broadcast in March 1986 and dedicated to the Giotto probe reaching Halley's Comet, which was to ignite his passion for these icy interlopers. As the years passed, he began writing astronomy articles for local news magazines before moving into internet radio where he initially guested on the Astronomyfm show 'Under British Skies', before becoming a co-host for a short time. In 2013 he created Comet Watch, a Facebook group dedicated to comets of the past, present and future. His involvement with Astronomyfm led to the creation of the monthly radio show Comet Watch, which is now in its fourth year. Neil lives in Suffolk with his partner and three children. Perhaps rather fittingly, given Neil's interest in asteroids, he has one named in his honour, this being the main belt asteroid 314650 Neilnorman, discovered in July 2006 by English amateur astronomer Matt Dawson.

Damian Peach is a world renowned and multi award winning astrophotographer. He has provided astronomical images for magazines and books including *Astronomy*, *Sky & Telescope*, *Astronomy Now* and *The Sky at Night* and has also appeared numerous times on BBC television promoting astronomy and astrophotography as well as giving talks on the subject both in the UK and overseas.

Damian was awarded both the British Astronomical Association Merlin Medal and the Association of Lunar and Planetary Observers Walter Haas Award for outstanding contributions to planetary astronomy. In 2009 he was part of a record-setting team that produced the largest ground based image mosaic of the Moon ever taken. He featured in the acclaimed national Explorers of the Universe photographic exhibition at the Royal Albert Hall and also had his work featured at the Edinburgh Science Festival. His work has also been used by NASA and ESA to illustrate the importance and quality of amateur planetary images. In 2011 he was crowned overall winner of the Royal Greenwich Observatory Astrophotographer of the Year competition as well as being a prize winning finalist in 2012 and 2013. He also recently won first place in both the National Science Foundation's Comet ISON photo competition and in the solar system category of the APOTY 2016 competition.

Katrin Raynor-Evans lives in Pontypridd, South Wales and is a Fellow of the Royal Astronomical Society and Royal Geographical Society. She is a prolific collector of astronomy and space stamps and is a member of the European Astronomical Society and Astro Space Stamp Society. In her spare time she writes articles and interviews for popular astronomy and philatelic magazines, including *Sky at Night* magazine, *Gibbons Stamp Monthly* and the American Philatelic Society journal *The American Philatelist*. She is also the Features Editor for the Society for Popular Astronomy's magazine *Popular Astronomy*.

Peter Rea has had a keen interest in lunar and planetary exploration since the early 1960s and frequently lectures on the subject. He helped found the Cleethorpes and District Astronomical Society in 1969. In April of 1972 he was at the Kennedy Space Centre in Florida to see the launch of Apollo 16 to the moon and in October 1997 was at the southern end of Cape Canaveral to see the launch of Cassini to Saturn. He would still like to see a total solar eclipse as the expedition he was on to see the 1973 eclipse in Mali had vehicle trouble and the meteorologists decided he was not going to see the 1999 eclipse from Devon. He lives in Lincolnshire with his wife Anne and has a daughter who resides in New Zealand.

Martin Rees is the UK's Astronomer Royal and is based at Cambridge University where he is a Fellow (and Former Master) of Trinity College. He is also a member of the House of Lords, and a former President of the Royal Society. His research interests include space exploration, black holes, galaxy formation, the multiverse and prospects for extraterrestrial life. He is co-founder of the Centre for the Study of Existential Risks at Cambridge University (CSER), and in addition to over 500 research papers he has written ten books and many general articles.

P. Clay Sherrod is the director and founder (in 1971) of Arkansas Sky Observatories - the oldest entirely privately-funded astronomical and earth science research facility in the USA. He is the author of 36 books in science and nature, as well as two novels, a book of poetry and a popular cookbook. Sherrod is known world-wide for his creative writings and lectures in the sciences and has contributed over five decades of research in astronomy, palaeontology, archaeology, environmental and earth sciences, and in the fields of molecular biology and research into the origin of life. His publications are detailed at **arksky.org/publications** and the Arkansas Sky Observatories can be accessed at **arksky.org**

Lynne Marie Stockman holds degrees in mathematics from Whitman College, the University of Washington and the University of London, and has studied astronomy at both undergraduate and postgraduate levels. She is a native of North Idaho but has lived in Britain since 1992. Lynne was an early pioneer of the World Wide Web: with her husband David Harper, she created the web site **obliquity.com** in 1998 to share their interest in astronomy, computing, family history and cats.

Susan Stubbs first became interested in stars as a child when she was astonished by the appearance of the Milky Way after seeing it from a dark sky site in Sussex whilst camping with the Guides. However, progressing to her Guide Stargazers badge showed her that not so many people were interested in astronomy back then, as finding an enthusiast to test her proved problematic! University, career and family then got in the way for quite a few years, but she returned to the fold via Open University astronomy courses and through joining Bradford Astronomical Society of which she is an active member. She was delighted to find that so many more people are interested in astronomy now than all those years ago in childhood. Susan recently had a new astronomical experience, travelling with her husband Robin to see the August 2017 total solar eclipse in Wyoming. She definitely wants to do it again!

Gary Yule developed an interest in astronomy at the age of seven when his father woke him in the middle of the night to observe Halley's Comet. Since then he has become a keen amateur astronomer and his journey has taken him down many avenues, ranging from the history of astronomy to astrophotography. Gary has a particular interest in antique telescopes and the stories behind them, and is the Chairperson and Curator of Instruments for Salford Astronomical Society where he dedicates most of his time. He is also involved in the organisation of the North West Astronomy Festival and heads up various astronomy and astrophotography pages on social media, in addition to which he buys and sells telescopes and mounts, many of which are antique.

2017 Yearbook of Astronomy

Editors RICHARD PEARSON, FRAS and BRIAN JONES

As many of you will be aware, the future of the Yearbook of Astronomy was under threat following the decision to make the 2016 edition the last. However, the series was rescued, both through the publication of a special 2017 edition (which successfully maintained the continuity of the Yearbook) and a successful search for a publisher to take this iconic publication on and to carry it to even greater heights.

The *Yearbook of Astronomy 2017* was a limited edition, although copies are still available to purchase. It should be borne in mind that you would not be obtaining the 2017 edition as a current guide to the night sky, but as the landmark edition of the *Yearbook of Astronomy* which fulfilled its purpose of keeping the series alive, and which heralded in the new generation of this highly valued and treasured publication. You can order your copy of the 2017 edition at **www.starlight-nights.co.uk/subscriber-2017-yearbook-astronomy**